# Online Intersex Communities

# ONLINE INTERSEX COMMUNITIES

## Virtual Neighborhoods of Support and Activism

*Brian Still*

CAMBRIA PRESS

AMHERST, NEW YORK

Requests for permission should be directed to:
permissions@cambriapress.com, or mailed to:
Cambria Press
20 Northpointe Parkway, Suite 188
Amherst, NY 14228

Library of Congress Cataloging-in-Publication Data

Still, Brian, 1968-
Online intersex communities : virtual neighborhoods of support and activism / Brian Still.
p. cm.
Includes bibliographical references and index.
ISBN 978-1-60497-592-5 (alk. paper)
1. Online social networks. 2. Intersex community. 3. Internet—Social aspects. 4. Intersexuality—Social aspects. I. Title.

HM742.S85 2008
306.76—dc22

2008048118

*To Amy, Jack, Olivia, and Abe*

# Table of Contents

# List of Figures and Table

# Preface

GLENDOWER: I can call spirits from the vasty deep.
HOTSPUR: Why, so can I, or so can any man, But will they come when you do call for them?
GLENDOWER: Why, I can teach you, cousin, to command the devil.
HOTSPUR: And I can teach thee, coz, to shame the devil By Telling Truth. Tell truth and shame the devil. If thou have power to raise him, bring him hither, And I'll be sworn I have power to shame him hence. O, while you live, tell truth and shame the devil!

(Shakespeare, *Henry IV* 3.1.52–3.1.61)

This book is about virtual intersex activism. It fills a gap in existing intersex research by focusing on how intersex persons, their families, and their advocates, through their ongoing construction of virtual communities, produce alternative rhetoric that challenges and, as recent trends show, contributes to the reshaping of the discourse surrounding intersexuality, its meanings, and its methods of medical treatment.

What is intersexuality? Once referred to as hermaphrodites, intersexed persons are born with a sexual anatomy or physiology inconsistent with social expectations of what constitutes a normal male or female. They may not have all of the required organs to be identified as a particular sex; a female, for example, may be born without a vagina, or a male without a penis. Others possess the parts of both sexes—one body with both testicles and ovaries, penis and clitoris. Finally, some children with all of the right parts for a specific sex are still deemed intersexed because those parts are sufficiently ambiguous to prevent doctors from quickly or easily determining sexual identity, or their appearance deviates too far from the norms for newborn girls and boys.

Of course, this definition of intersexuality, like all definitions, is rhetorically charged. "It is by our use of things," Stuart Hall writes in *Representation: Cultural Representations and Signifying Practices*, "and what we say, think and feel about them—how we represent them—that we give them meaning" (3). In other words, intersexuality does not have meaning in and of itself that can be separated from the culture in which it resides. It is a product not just of scientific fact but also of a myriad of cultural forces that have changed through time, and with it our perceptions of what is normal, of who should be corrected if they are not deemed normal, of when and how this treatment should take place, and of who has authority to speak on such things. As Emi Koyoma notes on the Intersex Initiative Web site, "There is no single 'intersex body'; it encompasses a wide variety of conditions that do not have anything in common except that they are deemed 'abnormal' by the society. What makes intersex people similar is their experiences of medicalization, not biology. Intersex is not an identity."

Until the last two decades, and in fact not uniformly until the last few years, those born with intersex conditions were treated quickly, usually in the first 18 months of life, long before they had enough knowledge and maturity to agree or dissent to treatment. Very often this treatment involved surgical conversion so that the "afflicted" intersex child could be made into a normal-looking male or female. Almost always the decision to do this was one that did not adequately involve parents or did not properly inform or educate them. Certainly the patient, the intersex child, had no say, and standard practice after treatment included covering up any knowledge of the treatment itself, meaning that many children grew up not knowing what had happened to them, not knowing that there were others like them. The result, therefore, for so many children born with an intersex condition was that they were subjected to a life of shame, of secrecy, and of enervating stigmatization that made them feel isolated and abnormal.

But a rise in intersex activism, fueled primarily by intersexed persons who have come of age, along with medical practitioners they had persuaded to become allies, has led to remarkable changes to the previous medical paradigm that had been subjected upon literally thousands of intersex children.

## Previous Literature

A great deal of important literature has been dedicated to the origins of this intersex activism as well as the history of the treatment that necessitated it. There is, for example, Alice Domurat Dreger's landmark book, *Hermaphrodites and the Medical Invention of Sex* (1998), which traces the growth in medical authority over the body, as well as the historical and cultural trends that led to intersexuality as an abnormality that required immediate medical and often surgical treatment. Anne Fausto-Sterling's *Sexing the Body*, appearing in 2000, offers insights into the cultural forces at work in defining sexual identity, as well as treating those instances of bodies that defy socially defined gender roles. Suzanne Kessler's *Lessons from the Intersexed* (2003) relies on interviews with medical practitioners as well as intersexed persons and their parents as she explores the boundaries of gender. In particular, she takes issue with the assumption behind the historical treatment of intersexed persons, which is the "emphasis on the genitals as evidence of gender" (7). Kessler argues for gender's constitution to be less defined along binary lines and more as something variable not with only two options but rather with varying degrees of possibility (9). Sharon Preves' *Intersex and Identity: The Contested Self* (2003) explores, through interviews with intersexed persons, how individuals cope with the demands of living abnormally, or being regarded as sexually deviant because of societal, not scientific, definitions. As she writes, "[B]odies that are considered normal or abnormal are not inherently that way. They are, rather, classified as aberrant or customary by social agreement" (3).

A number of articles have looked at the ethics of treatment regarding intersexed persons. Some, such as those by Dreger, Sharon Sytsma, Edmund Howe, Kenneth Zucker, and Iain Morland, appear in Sytsma's edited collection, *Ethics and Intersex* (2006). Milton Diamond, one of the earliest critics of the medical paradigm long in practice for treating intersexuality, has penned a number of important essays challenging the status quo, including pieces with Keith Sigmundson (1997) and Kenneth Kipnis (1999). Sheri Groveman, among others, has advocated for changes to methods of treatment (1999). Cheryl Chase, founder of the Intersex Society of North America (ISNA), has long been a voice for change. As an intersexed person, she has spoken from a position of credibility, arguing forcefully against stigmatizing methods of medicalizing intersexuality that were once commonplace (1999, 2003).

It is this rhetoric, along with the intersex activism it has helped to propel, that Preves examined in an influential essay, "Out of the O.R. and into the Streets: Exploring the Impact of Intersex Media Activism," that appears in the book, *Politics of Change: Sexuality, Gender, and Aging* (2004). Preves argues that the "intersex movement's use of mass media as a tool to reframe medical and lay conceptions" of intersexuality has enabled not only intersexed persons to reconsider what it means to be intersexed, but it has also motivated the medical establishment to question as well as modify their positions and the ways they treat intersexuality (180).

Preves' analysis of mass media's role in intersex activism does not place much focus on the Internet. However, she suggests that the "use of the Internet as a primary means of contact, education, and support has afforded the development of a geographically diverse advocacy movement" (188). Virtual communities dedicated to intersex issues have "given former patients and family members an opportunity to discuss intersex beyond the clinic" (188). So there is a need, she argues, for studying these communities further, to understand how such online media "clarify the processes through which categories of normalcy, deviance, health, and pathology are socially and politically constructed" (217).

But to date no such study of this kind has been carried out. This book remedies this by offering a focused, in-depth rhetorical analysis of the significant role virtual intersex communities have played in contributing to intersex activism. These electronic intersex communities or what the cultural anthropologist Arjun Appadurai calls "virtual neighborhoods" (195) are home to a wide range of intersex persons, their families, advocates, and even those beyond, attracted to the discussions they generate or the information they offer for professional reasons, or perhaps just curiosity. They enable their participants

- to learn about who they are and that they are not alone or "rare," whereas in the physical world, by contrast, their upbringing, physical appearance, and the enforcing norms of society often prevent them from doing so;
- to tell their stories, often for the first time, or have them heard by those that may be helped by them or may be able to offer help;
- to give and receive advice in an environment where they will not be stigmatized or shamed;
- to find a medical practitioner that is recommended by other intersex persons; and
- to advocate for change to established discourse, not as a single voice, but as part of an organized group of others committed to engaging traditional medical discourse using a variety of methods.

In this self-production, they not only produce meanings for themselves, but they also produce rhetoric that challenges and contributes to the redefinition of meanings beyond.

# Acknowledgments

I first wish to thank Toni Tan, who has patiently and expertly guided me as the book took form, went through revisions, and finally was made ready for publication. Along those lines, the feedback of peer reviewers, who offered critical input on how this book should be better, was much appreciated, and I owe a great debt.

This book, at least an early version of it, began in 2003 at the University of South Dakota where I worked with Susan Wolfe. She's been a friend and mentor for many years and her initial ideas, as well as revisions of my early attempts to formulate a thesis on the subject, were instrumental. You can't do anything without good teachers.

Amy Koerber read the book after its first round of review and her evaluation was extraordinarily useful, helping me to sharpen my argument especially as it related to the rhetorical and cultural analysis of the selected intersex Web sites. She also read a final version of the book to ensure that I had faithfully adhered to reviewer feedback and my language was correct and clear.

Finally, during research for the book, I conducted interviews with a number of leading intersex activists and scholars, including Alice Domurat Dreger, Betsy Driver, Cheryl Chase, Emi Koyoma, and Lynnell Stephani Long. Their passion is inspiring. All in their own right have had an important role to play in shaping and changing intersex discourse. They have stood up to make sure that the voices of so many others can be heard. I acknowledge them as much as I dedicate this book to them. By any means necessary, when often traditional means were not as easily available, they have critiqued the truth, and in that critiquing, online and off, they have changed what that truth means.

# Online Intersex Communities

# Chapter 1

# Introduction

Why a rhetorical analysis of intersex Web sites? Rhetoric, or "persuasion," Judy Segal argues, "is a central element in many medical situations" (1). She writes, "Experts persuade the public to count some states and behaviors as pathological and others as not" (1–2). Medicine, therefore, is not just the application of unerring science in an antiseptic, objective environment. Its constructions, its meanings, its truths, who it privileges to speak, and what can be said—all of it is contingent to the situation that such rhetoric is produced by and also helps to produce. To study the rhetoric of such a system, especially related to intersexuality, from the perspective of the rhetoric produced online, makes a great deal of sense. If rhetoric is the art of persuasion, then rhetorical analysis is an examination of how such persuasion is created, what influences it, and what it, in turn, influences.

To conduct a rhetorical analysis, then, is to conduct an analysis of the culture, of the discourse, that surrounds the words, the norms, the systems that shape, in this case, past medical practice that brought about a rise in intersex activism, and a resulting alternative rhetoric that has served to modify perception and practice of the treatment of intersexuality. Rhetorical analysis, therefore, is not just an amusing dance with words. Rhetoric is representative of and contributor to a complex, dynamic discourse of values, truths, and rules. A study of the rhetoric produced online by intersex communities, in response to a particular rhetoric espoused for some time regarding the treatment of intersexuality, enables us to take a different approach from previous criticism to comprehending the medicalization of intersexuality as well as the forces at play that have shaped productive resistance to it.

There are many examples of rhetoric at work as producer and product of the medical discourse surrounding intersexuality. Let's explore one briefly to show just how much intersexuality is a cultural or rhetorical construct. Read first Gayle's story, told online at the Androgen Insensitivity Syndrome's support group Web site:

> My daughter was born by C-Section. The gynecologist told us that we had a beautiful baby girl. Then she was taken by the pediatrician and was checked out. At that time I knew that something was wrong. It seemed to go quiet and people were gathering around her. I had been worried during my pregnancy because I had to take medication for my asthma, and one of the side effects of one of the medications, was that babies could be born with a cleft palate. So I thought that our baby had a cleft palate. The pediatrician then came over and told us that he wasn't sure if we had a boy or girl and that arrangements needed to be made to transfer the baby about seventy miles away

> to a hospital in Philadelphia. Apparently she needed to have a lot of tests done. To say that we were stunned is an understatement. The pediatrician tried to explain about ambiguous genitalia. He also said that she would probably be brought up female since it is harder, "to make a pole than a hole."
>
> ...After about three days of various tests, including a chromosome test which came back as male (XY) we were informed that it would be better to raise our baby as a female. So at that point we informed people that we had a baby girl. At three months old our baby was back in the hospital having her testes removed because according to the doctors they could become cancerous. Also she needed to have surgery done on her genitals. All through this we were still in shock. We didn't know of anyone else who had gone through this. Our pediatrician had only seen one other case of ambiguous genitalia and that was years ago.
>
> So we felt alone and isolated dealing with it all. We have told only a few people about her genetic make-up. Our biggest fear was that people would look at her and treat her differently. Would they look at her as a boy pretending to be a girl? Some people can be so cruel and as a mother I want to protect my daughter. I am not sure that what I am doing is right, all I know is that I love my daughter with all my heart and I am trying to do what is best for her.

Many doctors, such as the one that treated Gayle's child, once took what Fausto-Sterling has characterized as a phallocentric approach to decisions regarding intersexuality. Motivated by cultural norms, not medical reasons, doctors often converted boys to female because their penises were not considered viable enough to let the boys do what men should be able to do—urinate easily standing up or penetrate a vagina in a pleasurable way for the male and his partner (57). Since many surgeons felt that they weren't "very good," as Gayle suggests previously, "at creating the big, strong penis they [society] require men to have," female conversion was carried out for those boys that did not measure up (Fausto-Sterling 59). Writing in 1997, Lori Sherman, a practicing endocrinologist, reveals this cultural influence when she states in a matter-of-fact way that regardless of whether the child has an XY (male) chromosomal pair, "[I]f the phallus is deemed inadequate to assure normal sexual function…then a female gender assignment should be made with eventual removal of the gonads and construction of female external genitalia" (201).

Such a position, one that governed surgical treatment of children born too close to the line between normal and abnormal, provoked harsh responses from intersex activists, and a corresponding abundance of rhetoric meant to counter it in the hopes of changing it. The problem intersex activists see in doctors operating on children with intersex conditions is not necessarily that something is done when it should not be done. In some cases, such as hypospadias, when the urethral opening is awkwardly placed or the penis is too small for any sort of future sexual reproduction, surgery to lengthen the penis, as well as testosterone treatments to promote growth, does seem like a viable option. But the problem is that many past decisions to operate to surgically correct intersex "abnormalities" were not motivated solely on the basis of medical need. The fact is that many children declared abnormal and treated as such are not technically intersexed, nor are their conditions life-threatening so as to require emergency measures. They were operated on quite frequently, then, without the consent of the patient because parents wanted them to be normalized and because the prevailing discourse dictated that such normalization should occur. The discourse's rules, not medical need, therefore, required doctors to enforce normality. Prevailing rhetoric, therefore, not medical necessity, required surgical adjustment. As proof of this, John Money, the father of the medical paradigm once accepted without question for the treatment of intersexuality, wrote, along with co-author Patricia Tucker, in the book, *Sexual Signatures*, that "at the minimum extreme, an erect penis must be something over two and a half inches in length to penetrate far enough into a vagina for a man to begin to feel satisfied with what he can do for his partner…not being able to reciprocate in coitus can inflict terrible wounds on his ego" (56).

Despite the fact that genitals vary a great deal, the medical discourse surrounding intersexuality had in place for some time a measurement for normal based on cultural norms, and enforced by rhetoric, that rarely had anything to do with ensuring the physical well-being of the patient, aside from ego. As a result, in many cases those children otherwise healthy but perhaps with a clitoris too large or a penis too small were diagnosed as intersexual and then repaired or "normalized." Perception, therefore, driven by discursive norms, was as much a factor as chromosomes.

One example to further illustrate this can be found in the treatment of 5-Alpha-Reductase (5-AR) Deficiency, made infamous by Jeffrey Eugenides' 2002 novel, *Middlesex*. 5-AR affects only a limited population, one in which incest seems to be highly prevalent, and it has been documented as occurring quite frequently among some living in the Dominican Republic and among the Sambia people of New Guinea (Herdt 68). In this condition, children are born male, having both an XY chromosomal makeup and testicular gonad tissue. However, they lack the enzyme 5-alpha-reductase, so their bodies cannot convert testosterone into the dihydrotesterone necessary to create male genitalia. As a result, they are born with female-looking genitalia, including what appears to be "a short vagina and apparent labia and clitoris" (Dreger, *Hermaphrodites* 39).

But at puberty these children's testicles begin to produce additional testosterone, and the 5-AR enyzme is not required at that time to convert that testosterone properly. So roughly at 12 or 13 years of age 5-AR children, perceived to be female, become very clearly male. As Dreger notes, "[T]he body grows taller, stronger, more muscular…testes descend…and the penis/clitoris grows to look and act more like a penis" (*Hermaphrodites* 39).

Because this intersex condition has always occurred in the Sambian culture, where medical technology is limited, those born with this condition are not treated. In fact, they occupy a special place within the culture. For them, as Gilbert Herdt writes, the Sambia "have created an historically divergent third sex category, neither male or female" (420). In fact, the Sambia call 5-AR people *kwolu-aatmwol*, or "female thing changed into a male" (Herdt 68). In the Dominican Republic 5-AR also occurs so often that prevailing rhetoric has generated a name for those that are neither male nor female. 5-AR children not fully developed are called *guevedoche*, or "penis at 12" (Herdt 68).

Clearly, therefore, cultural biases and the discourses that reflect and maintain them determine what is considered sexually ambiguous or not, acceptable or not, normal or abnormal. Why? According to Alice Dreger, a person who does not properly perform that which is considered normal, such as the hermaphrodite or intersex person,

> continues to be a person whose body gets caught up in cultural 'border wars'—wars over the borders separating males and females, men and women, boys and girls, borders separating those with authority from those without. 'Doubts' inherent in the hermaphrodite's body are still

> exorcised and shrouded in silence, lest that doubtfulness leak and spread onto the child's identity, the parents and family, medicine and science, and finally the streets…We still insist there be no hermaphrodites, lest they grow and multiply. (*Hermaphrodites* 198)

Kessler and Wendy McKenna, in their study of culture and gender, laid out a number of expectations that society has regarding gender, all of which the intersexed person disrupts:

1. There are two, and only two, genders.
2. One's gender is invariant.
3. Genitals are the essential sign of gender.
4. Any exceptions to the two genders are not to be taken seriously.
5. There are no transfers from one gender to another except ceremonial ones.
6. Everyone must be classified as a member of one gender or the other.
7. The female/male dichotomy is a natural one.
8. Membership in one gender or the other is natural. (Preves 17)

Assuming that these expectations constitute the ideology behind our culture's discourses surrounding gender, sexual identity, and intersexuality, it is easy to understand how intersexuality, by its very existence, constitutes a kind of social deviance that must be considered far worse than other conditions or behaviors, such as public nudity, that are also regarded as unacceptable. For although Americans may look upon public nudity with shock and react in ways that tell the violators that their behavior is not acceptable, we react to intersexuality in a far more systematic, authoritative, and conclusive way. So severe is the deviation the intersexed person's body perpetrates that we alter that person's body, and his or her sexual identity, if need be. No other deviation spurs such a reaction, but because intersexuality represents such a deviation from a set of culturally imbued gender expectations considered inviolable, we have, according to Preves, "developed institutional means of covering it up or erasing the violation" (*Intersex and Identity* 20).

We see then that sexual identity, Thomas Laquer writes, "is situated; it is explicable only within the context of the battle over gender and power" (11). Beliefs about gender are tied to the discourses that produced them. Therefore, discourse, as well as the rhetoric that works within it that propagates its rules and leads to the modification of them, contextualizes and produces our knowledge of intersexuality.

Because of this, for Foucault, there is no true sex, no transcendent concept of gender. Gender derives its meaning, and we derive our knowledge of that meaning from "relations between institutions, economics and social processes, behavioral patterns, systems of norms" (*Archaeology* 45). In other words, what we think we know and believe about gender and sex is grounded in the discourses that produce them, and those discourses are, in turn, produced by the processes, patterns, theories, and conceptions derived from different disciplines, from historical and social formations, and from events.

Truth is shaped by discourses, which are shaped by the schools, the books, the media, the doctors, and the greater mass of people in general. That which is not the truth is in binary opposition and must be false, as long as the discourse's "will to truth" supports and thus gives veracity to that object of knowledge considered to be true. Once the truth has been established, discourse functions constantly to reinforce it as the truth. It establishes norms and produces rhetoric to regulate and reinforce it so that the individual knows how to behave in ways defined as acceptable by knowledge or truth. Those who cannot do so are punished for their transgressions. In medicine, for example, the intersex body was disciplined for a long time because it transgressed against the normal, defying what the normal body is supposed to be. Likewise, for a long time the voices of intersex activists were marginalized because their views were not normal, not in line with a truth that the discourse sought to reproduce through, in large part, the rhetoric of medical studies, of publications intended for the mainstream as well as more specialized readers, and even the education of future medical practitioners as well as new parents of intersex infants.

Of course, it is possible to change that which is regarded as truth. Intersex activists are engaged in trying to do this, have done this recently to a certain degree. But the will to truth is, Foucault writes, "reinforced and accompanied by whole strata of practices" (*Archaeology* 219), making it difficult to undermine the truth that it supports with an alternative it rejects. So the nature of power makes it difficult to question its truths because our subjectivity and even our perceptions of that subjectivity are the effects of what Foucault calls "'local centers' of power-knowledge" (*History of Sexuality*, 1: 98). Like the baby, all individuals are surrounded by various caregivers, are always under surveillance and subjected to various controlling techniques that transmit and enforce knowledge. In other words, power working from multiple, local centers, is omnipresent, shaping the thoughts and behaviors of all individuals.

For Foucault, discourse and power do not constitute a pyramid of control in which power and meaning occupy the top. Rather, power and meaning emanate from everyone and from everywhere, creating a web of all-encompassing forces so ubiquitous that it is virtually invisible. Thus, in the case of intersex children, parents once readily acceded with doctors' recommendations for treatment not only because they accepted the authority of the doctors but also because they shared the doctors' views on normalcy, which reflected those of society generally. This kind of power seen at work in the parents' decision Carolyn Dean calls "monstrous precisely because it can no longer be anchored, located, or contained; it now infuses, occupies, and produces everything" (280). A diffuse set of forces affects the parents. As Foucault argues, it is within the subject, whether it be the parent, the doctor, or the intersex child, that power can best be observed. In the *History of Sexuality, Volume I*, Foucault asserts that "modern societies" have not "governed sexuality through law and sovereignty" (90). Rather, they have relied on certain technologies, such as medicine, pedagogy, psychiatry, demography, media, and now, he might have said, even the Internet. The technologies provide concepts and norms for regulation that people absorb, engage with, and disseminate as subjects, because of and in reaction to power. These technologies make possible the construction of sex, just as they make possible the construction of personal identity.

But they also make possible the construction of alternative rhetoric that counters such truths. Judith Butler argues that gender is, like so many things, a discursive construct "performatively produced" (*Gender Trouble* 24). This means that if gender is performatively constructed, then any transgressions against normal gender, such as stories that run counter to official stories and truths, provide a rhetorical mechanism for changing discourses about gender and sex. In other words, because gender is continuously instantiated through individual performances of it, alternative performances, such as those told online, arguably have the potential to shift our concepts of gender and identity.

Of course, arguing that the potential exists to do that is easier than proving that it has happened. I am not equipped with the requisite data, nor do I attempt to make such a case in this book. But what I do try to point out is that intersex communities that did not or could not exist offline as they do now online are productive sources of alternative rhetoric. Because society values (perhaps now more than before) the words of "wounded storytellers," or those who do not treat but have been treated (and in many cases traumatized), then the intersex activist can speak with a measure of newfound authority. The Internet also represents what Paul du Gay et al. describes as "new media" (22), a technology that provides what might not have been easily accessible before, which is namely "the sources and the channels for the circulation of meanings within the culture" (23), impacting in the process how and who influences the production of culture. The Internet, like other new media such as the cell phone, "has opened up a new frontier in modern cultural life," du Gay et al. writes, "and completely transformed the process of 'meaning-making' which…is at the heart of culture" (23). As proof of this, Alice Dreger points out that since the Intersex Society of North America (ISNA) has made available online (as a downloadable PDF) the document "Guiding Tips for Parents," it has been downloaded more than three thousand times ("Question"). Parents, therefore, faced with the task of rearing intersex children are not turning solely to doctors for information. Rather, they are also seeking out information from those whose knowledge is based on lived experience.

It may be that all virtual intersex activism is not as purposeful as this ISNA parenting guide. In other words, it is possible that intersex activists may challenge doctors with alternative paradigms, issue press releases, and even declare intersex awareness or solidarity days, but such directly confrontational acts are not the only ways to resist. Just telling a story or responding to it, engaging in communication not possible before, offering advice for dealing with a medical issue or which doctor one should choose—these rhetorical acts are also examples of resistance because they are made by those who previously were not privileged to make them, and as they occur, they contribute to the production of a virtual community that, although influenced by discourse, is an environment that is, to a certain degree, beyond the control of the clinic and, thus, the authority of medical discourse. From such environments, virtual communities are formed that produce narratives of alternative rhetoric that break the rules or contribute to the redefinition of existing ones. Narratives, in fact, that, according to Segal, "have found a place in health research" (59) or are considered productive parts of the discourse that surrounds intersexuality, and, as a result, shapes what it means, and how treatment in response to it is carried out.

## METHODOLOGY AND OVERVIEW

To carry out a detailed examination of the nature of these virtual rhetorical narratives and the impact they've had in promoting intersex activism, I've divided the book into seven chapters including this introduction and a brief conclusion. The second chapter provides an overview of the rhetoric surrounding the medicalization of intersexuality (which is the primary motivator for intersex activism) by framing discussion around a 2000 American Academy of Pediatrics' document, and its revision coming just six years later, that provided recommendations for the treatment of intersexuality. The dramatic disparity in language and rhetoric in the two versions of the same guidelines signals the challenging power of intersex activism, just as it offers an insightful backdrop for a discussion of the cultural and historical issues that led to the creation of the first version of the document and as well as to the activism that initiated a drastic revision of it in 2006.

In chapter 3, I define *virtual community* as a concept. I also apply Constance Porter's useful typology of virtual communities to distinguish the different types of intersex Web sites. Then I use Appadurai's idea of virtual neighborhoods, along with critical input from Paul du Gay's circular conceptualization of cultural production, to lay the theoretical foundation applied in the final chapters of the book as a means of assessing the rhetoric produced online by intersex Web sites. Finally, in the three analysis chapters, coming before the conclusion, I dedicate each one to an examination of the three significant kinds of rhetorical constructs, bonding, disseminating, and confronting, that serve to build productive intersex virtual neighborhoods:

- **Bonding (chapter 4):** Just by going online and meeting with each other, by telling stories, by listening and responding, intersexed persons create change because they defy their previous isolation and they form resistant communities that had not existed before of people like them or who believe in what they believe.
- **Disseminating (chapter 5):** Previously, intersex activists had limited means to disseminate information that challenged their treatment—intersex Web sites afford them the opportunity to produce easily accessible, constantly available information.
- **Confronting (chapter 6):** Offering press releases advocating for different methods of treatment, providing patient information, for example, to parents of children diagnosed as intersexed, recommending medical practitioners—all

> of these confront established authority on intersexuality because such information isn't coming from doctors, but instead is coming from intersexed persons themselves.

The Web sites I have selected for my examination are a self-selected, cross-section of intersex sites. Accordingly, I look at some that represent different intersex conditions (such as Androgen Insensitivity Syndrome or Turner's Syndrome), assuming that how they constitute themselves may be different because of their special focus. I also look at those that allow different levels of community interaction. Those sites, for example, that lack a discussion forum or other asynchronous, or even synchronous (such as instant messaging), communication tools will not be able to provide the same sense of community that more technologically advanced sites afford their participants. Additionally, I look carefully at those intersex Web sites that consider themselves, or are considered by intersex persons, to be dedicated to issues affecting the entire intersex community. Consequently, sites like Bodies Like Ours (BLO), Organisation Intersex Internationale (OII), Intersex Initiative, and the Intersex Society of North America (ISNA) are discussed frequently. Finally, for ethical reasons, I have chosen not to use discussion postings that require membership and user login to access, since the assumption I've made is that the information contained in them is private and not meant, therefore, for public use and study. However, postings not requiring login access are used as examples throughout, although I have elected to change the original pseudonyms of posters as a way to protect their identities.

Precedent for this can be found in similar analyses of online discussion forums. Denise Carter, for example, in looking at discussions at Cybercity, a virtual community, suggests, "while using nicknames on the Internet might be seen as automatically protecting one's own identity, this is not necessarily the case" (153). In fact, Carter writes that, given the "range and accessibility of information on the Internet," a nickname or pseudonym "is a very specific identity" (153). Sorin Matei addresses this in research he conducted of conversations found at the Well, an online discussion forum, by replacing pseudonyms with letter combinations, such as xxx ("From Counterculture"). For my examination, I have assigned different pseudonyms, replacing, as an example, "Kathy" with "Sue." I have also chosen not to include the URL of the discussion post thread in my list of references as a final means of ensuring that the poster's identity, even if it is a nickname, is protected. Intersex persons have been stigmatized enough. I want to analyze their online communication, but not in a way that further exploits them.

My examination will show that what these sites and so many others represent, from personal blogs to e-mail lists, is the opportunity for those isolated by the nature of their treatment and the geographic distance that often stands between them, and others like them, to feel a sense of belonging, to exchange information, and to contribute in big and small ways to the production of counter-discourse. What kind of rhetoric do they generate that furthers intersex activism? Can they summon spirits? Can they tell truths to shame the devil? I say yes. The analysis to come will show how.

# Chapter 2

# The Rhetoric of Intersexuality

## 2000 American Academy of Pediatrics Statement

In July 2000 the American Academy of Pediatrics' Committee on Genetics issued a consensus statement providing recommendations for the evaluation and treatment of those children born intersexual, or, as the statement suggests, "whose genitalia are obviously ambiguous" at birth (139). The "abnormal" intersex child, the 2000 statement reads, "constitutes a social emergency" (138). As a result, the child should be treated as soon as possible, and the abnormality should be corrected, even if it requires surgical conversion of the child's perceived ambiguous sex to one more "normal," so that the child will be normalized, parent's anxieties will be lessened and the "social emergency" will go away.

Rhetorically, this is a powerful document, possessing an ideal mix of the three modes necessary according to Aristotle for persuasion: ethos, pathos, and logos. Ethos is about credibility, or how speakers convince us that they are authorities on the subject. Certainly, the fact that the authors of the AAP Statement are all medical doctors lends credibility to their argument. As does the location for the rhetorical act, which is the most highly regarded, peer-reviewed journal for pediatrics, the one that all others seek to emulate, through which derivative synopses in various forms (press releases, TV/radio news features, pamphlets, medical textbooks) flow to other audiences. Also, it is the touchstone document that practitioners in the field adhere to when making medical decisions, such as treating intersexuality.

Pathos, or the emotional appeal to an audience, varies greatly depending on the situation. It need not be something over-the-top, making an audience cry because it is so passionate. Pathos is about making a connection, reaching out to more than just an audience's logic. There is already a strong built-in pathos between writers and audience for this consensus statement because they hold shared values and experiences as medical practitioners tasked with treating, primarily, children. Of course, the document does make emotional appeals, attempting to spur its audience to regard cases of sexual ambiguity at birth as not typical but, as the document suggests, "emergencies" (138). The word *abnormality* is also used frequently, again suggesting by way of emotional appeal that something is wrong that must be corrected.

Finally, this document has rhetorical or persuasive force because of its appeal to logos, or its use of facts and figures. It adheres to a method of argument that is acceptable in this environment and by its intended audience. It presents a definition for sexual ambiguity, a method for conducting a clinical history and examination, a background on the causes for ambiguous genitalia, and a collection of factors to be used for deciding the corrected sex of the child. The presentation of the facts, including charting the possible causes and laying out in detail reasons to consider for determining a new sex, appeal logically to the audience in question because it appears to be thorough, and because it is based upon already accepted ideas or truths that pervade the prevailing discourse surrounding the treatment of intersexuality.

In other words, what makes this such a powerful rhetorical document is that it is a product of the existing system. It conforms to system expectations, it comes from the most privileged speakers of the discourse, it appears in a highly regarded communicative deliverable, and it reflects, at least at the time it was published in 2000, an approach to the classification and treatment of intersex children that had been widely accepted and enforced since the late nineteenth century.

## Intersex Treatment Origins

Before this time, even when doctors had access and authority to treat hermaphrodites, they did not seek to correct them. But, as Elizabeth Reis' insightful historical essay on intersex persons in the United States argues, a growing concern over the boundaries of gender led many physicians to adopt new approaches to dealing with the ambiguity regarding it. "Whether it was to ensure the legal status of men or women or to show that sex, like race, should be something uncomplicated, permanent, and easy to determine," Reis writes, "nineteenth-century doctors insisted on certainty rather than ambiguity in gender designation" (para. 6). Confirming this is language from a nineteenth-century physician, Blair Bell: "'[T]he possession of a [single] sex is a necessity of our social order, for hermaphrodites as well as for normal subjects'" (Dreger, *Hermaphrodites* 161).

Fueling such insistence on there being possible only two clearly defined and separable sexes was the connection that at least doctors in the United States made between hermaphroditism and homosexuality. As Christina Matta points out, "[H]omosexuality and hermaphroditism—representing the moral and physical sides of sexuality respectively—were both deviant conditions. The sexual pervert, regardless of the exact nature of his or her perversity, was a trick of nature, and stood in clear contrast to normal sexuality" (80). The result, therefore, was that hermaphrodites increasingly became no longer just fascinating objects for study. They were not even considered viable outcasts on the edges of society. Rather, they represented in their bodies a kind of "sexual depravity" (78) that could not, like homosexuality, be tolerated. Accordingly, before 1879, Matta explains, only three times in forty-one published medical reports was surgery recommended as an option for treating hermaphroditism. But after that period, it was suggested with far greater regularity, and, by 1904, was considered the most justifiable approach (81–82). By making the hermaphrodites either clear men or women, the hermaphrodites would be cured of deviant sexual practices and their bodies would no longer exist as challenges to now firmly rooted beliefs about gender.

Dreger, writing in *Hermaphrodites and the Medical Invention of Sex*, calls this period that Matta documents previously the "Age of Gonads" (139). In 1876 Theodore Klebs, a scientist later noted for discovering the bacteria that causes diphtheria, postulated that a person's "true sex should be based exclusively on the nature of the gonads" (145). Klebs argued that regardless of what a person appeared to be or not to be, that person had a true sex and that sex could be discerned by careful examination of the gonads (146). "All others with mixed anatomics," Fausto-Sterling writes, "no longer...qualified as true hermaphrodites" (38). In other words, there was one, and only one, possible sex for every person, and their genitalia did not lie (Dreger, *Hermaphrodites* 109). From that point on, there were almost no more cases of hermaphrodites. In fact, drawing on Kleb's position, George Blackler and William Lawrence wrote a paper in 1896 that examined earlier claims of true hermaphrodites. Based on their findings, they concluded that, of the 24, only three could be considered "true" hermaphrodites. As a result, the previous existence of people living, in a sense, as both sexes was removed. As Fausto-Sterling writes, doctors like Blackler and Lawrence, in enforcing the new medical position on sexuality, "cleansed past records of hermaphrodites, claiming they did not meet scientific standards" (39). There might be a male with too small a penis or a woman with too large a clitoris, and both might be considered monsters of nature, but one was male and the other female—there were no other possibilities. As Dreger points out, the hermaphrodite was no longer a double-sexed person; instead, "[T]he hermaphrodite was just a would-be male or female gone wrong in the womb" (*Hermaphrodites* 34).

Because the doctor alone had the training and equipment necessary to view gonadal tissue and determine true sexual identity, the authority of the doctor in all matters concerning sexuality became preeminent (Dreger, *Hermaphrodites* 61). By the 1930s, the Age of the Gonads gave way to what Fausto-Sterling calls the "Age of Conversion" (40). Doctors at that point no longer just had the power to name. They also had the power to intervene surgically, and many did because, as Fausto-Sterling asserts, they "found it imperative to catch mixed-sex people at birth and convert them, by any means necessary, to either male or female" (40). As this trend continued through the twentieth century, the term *hermaphrodite* was replaced with *intersexuality*, a term more reflective of thinking in society and in medicine that regarded those bodies in between as abnormal.

## JOHN MONEY'S MEDICAL PARADIGM

Wrapped around this new description was a paradigm for treatment that the 2000 AAP Consensus Statement reflects in its language. The paradigm was given a great deal of its rhetorical charge through the work of John Money, a psychologist who spent most of his career at Johns Hopkins University. Based on a select few cases, Money, along with other sexologists, hypothesized that any child born, regardless of anatomy, is basically "gender neutral." This means that the child can be influenced, through professional counseling and proper upbringing, to be either a male or female. Nurture, therefore, is more responsible for sexual identity than nature. As William Reiner points out, Money's approach, as seen detailed in the following principles, takes for granted that "an appropriately 'engineered' environment" can shape the child into the normal male or female identity it has been assigned ("Gender" 177):

1. People essentially have no sexuality at birth; their chromosomes and their genitalia are not as important as how they are raised, or how society shapes them (Crouch 31).
2. A child can be made into either a male or a female, but the key is that the child is never informed of his/her past and is raised unambiguously (Preves, *Intersex* 53).
3. Action must be taken quickly (it is an emergency, the 2000 AAP Statement points out) to fit the child, surgically, psychologically, socially, into the assigned sexual identity. Basically, until we learn to talk and therefore tell, understand, and remember stories and thus digest

cultural practices, we can be shaped. After that point it will be too late except for a very few. So Money suggests that sexual identity needs to be assigned before 18 months of age (*Sexual Signatures* 98).

4. People must have normal-looking genitalia to be able to adapt to the sexual identity given to them. If they do not see themselves as a normal male or female, then they will not psychologically accept themselves as that assigned sexual identity. Furthermore, parental anxiety will be heightened and parents will not actively nurture the gender-assigned child if they do not see what they believe to be a normal-looking male or female (Preves, *Intersex* 53).
5. Finally, for the child to be healthy, the parents must agree with the treating medical professionals to keep secret what has happened, and even to lie if necessary, so that the child will believe he/she has always been a normal male or female (Preves, *Intersex* 54).

Normal-looking genitalia, along with the ability to function normally in sexual terms, are absolutely necessary for a child to behave normally. Money writes,

> The closer the resemblance of the external organs to the normal female, the less the likelihood that anything can ever be done to permit a normal male sexual life. Therefore there are many male hermaphrodites who should be designated female at birth, with a program then…of appropriate surgical correction and hormonal treatment. So treated, and with the parents properly counseled, such individuals grow up and function very well as girls. (*Sex Errors* 144)

"Money's theory and recommendations," Jorge Daaboul and Joel Frader write in 2001, "have guided professional practice for the last 40 years and can be found in current, standard pediatric endocrinology textbooks" (1577). One well-respected pediatric endocrinologist, Mel Grumbach, asserted in the same year that Money "was decisive in the universal acceptance not only of the theory that human beings are psychosexually malleable at birth, but also of sex reassignment surgery" (Colapinto 75). For a long time then, Money's influence was seen at work in a medical paradigm that was generally regarded as the best approach for defining and treating intersexuality.

Of course, there were objections to it. As early as the 1950s, Diamond and other medical practitioners questioned Money's theories (Colapinto 45). In the 1970s Bernard Zuger, a psychiatrist, published a study that found several adolescents had, in direct contradiction to Money's theory, rejected their assigned sexual identity (Fausto-Sterling 68). Still, such alternative voices were, by and large, ignored inside the medical community and in the greater spaces of the surrounding discourse. Why? Again, Foucault's concept of the "will to truth," as well as the power of the rhetoric, is useful for understanding. If Money's ideas and the paradigm for treating intersexuality we see in the 2000 AAP Statement are enforced everywhere, in medical textbooks, in highly regarded medical literature, and in a variety of mainstream publications, then the rhetoric produced for multiple audiences, in different formats, and at different levels of understanding serves to instantiate the truth while it makes difficult to be heard the voice of opposition to that truth.

Ironically, at one time, doctors, scientists, jurists, church leaders, others in positions of authority, and society by and large accepted the idea that nature created variety in human sexuality. In fact, as far back as the Tosefta, an early Jewish religious text (third century AD), we see provisions made for the hermaphrodite. For example, hermaphrodites were forbidden to shave. More important, they also could not inherit the estates of their fathers (Fausto-Sterling 33). Early biblical scholars even debated, based on their readings of *Genesis*, whether Adam was a hermaphrodite, half of whom was split off from him to make a woman, Eve (Reizenberger).

By the Middle Ages in France, the canonical and civil courts gave the power of declaring the sexual identity of hermaphrodites to them and their guardians. At baptism, the father or godfather announced what the sex was, and then, later in life, if the hermaphrodite chose to marry, he/she could confirm that previous designation or choose the other sex. The only restriction, according to Foucault, was that the hermaphrodite could not change back after making his/her selection at marriage (*Herculine* viii). Finally, as Matta points out, doctors were studying hermaphrodites closely up into the middle half of the nineteenth century, submitting case studies of their examinations for inclusion in journals. But a shift in the discourse changed all that and, quite literally, within a period of just a few short years, there were no reportings of hermaphrodites. New truths were put into place about normality and abnormality, about what it means to be, to look like, a man and a woman, and about what should be done and how it should be done when children are born who defy expectations.

But almost as suddenly, certainly as dramatically, the will to truth that had propelled so many surgical conversions of children born "in-between," that the 2000 AAP Statement represented and empowered, was called into question—in fact, was challenged and even modified. Therefore, when the AAP issued another consensus statement on the recommended treatment of intersexed children in 2006, only six years after the 2000 statement, a new rhetoric, reflective of new truths, was revealed.

## 2006 American Academy of Pediatrics Statement—A Comparison

We need only look comparatively at the first sentence in both statements to see the change in tone. In 2000 the statement begins, "The birth of a child with ambiguous genitalia constitutes a social emergency" (138). The 2006 Statement, published in the same journal with Peter Lee as lead author, and also published in the *Archives of Diseases of Childhood* with I.A. Hughes as lead author, is meant to be a revision of the 2000 Statement. Its first sentence reads as follows: "The birth of an intersex child prompts a long-term management strategy that involves myriad professionals working with the family" (488). We see no mention of emergency. Rather, patience is suggested in the use of "long-term." No treatment of the child as something amorphous or ambiguous. The child has a status, is intersexed. Or rather, the child's essence is not defined as abnormal. In fact, new language suggests even the term intersex be replaced with the phrase "Disorders of Sexual Development" or DSD (488). The child "is" not something but instead has something.

A side-by-side comparison of other language in the two recommendations published just six years apart in the same journal also illustrate the seismic shift in rhetoric (see table 1).

Note that harshness gives way from 2000 to 2006. The child is not an abnormality to be corrected but has potential to prosper, to adjust. With regard to the decisions involved in determining the sex of rearing, family views, as well as what the 2006 document calls "circumstances relating to cultural practices" (491), are now valued as much as physiological factors, such as endocrine function, that were only considered in 2000. It is not that the 2006 Statement recommends for the creation of a third sex. A child still should be assigned to a sex. But how that is done, how the child is regarded, and what influences the

**TABLE 1.** Comparison of AAP statements on the treatment of intersexed children.

| **2000 AAP Evaluation** | **2006 Consensus Statement (all page numbers from Lee et al., *Pediatrics*)** |
|---|---|
| "[A]bnormal appearance can be corrected and the child raised as a boy or a girl as appropriate" (138). | "A key point to emphasize is that the child with a DSD has the potential to become a well-adjusted, functional member of society" (490). |
| "As much as possible, the parents need to be included in the discussions regarding sex of rearing…The emotional tone established by healthcare professionals in conveying this information can have a lasting influence on how the parents conceptualize the abnormal genital development" (138). | "Parents need to be informed about sexual development, and Web-based information may be helpful" (490). |
| "ambiguous genitalia" (used on several occasions from 138–141). | "Terms such as 'intersex,' 'pseudohermaphroditism,' 'hermaphroditism,' 'sex reversal,' and gender-biased diagnostic labels are particularly controversial. These terms are perceived as potentially pejorative by patients and can be confusing to practitioners and parents alike. We propose the term disorders of sex development (DSD)" (488). |
| "When problems are apparent, help should be sought from psychologists, counselors, and other mental health professionals with experience in intersex disorders" (141). | "The value of peer and parent support for many chronic medical conditions is widely accepted, and DSD, being lifelong conditions that affect developmental tasks at many stages of life, are no exception…<br>• Peer support ends isolation and stigma, providing a context in which conditions are put into perspective and intimate issues of concern can be discussed safely with someone who has "been there" (496).<br>"Support groups complement the work of the health care team and, together, can help improve services" (496). |

decision-making process has changed. Even who has a say in the process has changed. There is privileging of not just new truths but also new authorities that can speak the truth. In 2000 doctors or mental health professionals are the main actors. In 2006 peer support beyond the clinic, even found online, is regarded as useful to intersex children and their parents. Parents also are treated less as laypersons to be educated and instructed, but instead as important stakeholders (the most important next to the children) who must be counseled, must be assisted in the decision they, not doctors, will make.

Why such a change in rhetoric, in the truth within the discourse that such rhetoric espouses and enforces? Why are now intersex activists like Cheryl Chase contributors to a 2006 statement when only six years before they could only be the ambiguous subjects of such a statement?

Any shift from one paradigm of thinking to another, such as that which intersex activists' advocate, is messy and, as Thomas Kuhn notes, "never done overnight" (7). After all, new ideas, such as those that intersex activists espouse, threaten "a change in the rules," according to Kuhn, "governing the prior practices of science" (7). These rules, based upon much research and application, represent a fundamental way of thinking, considered tested, proven, and thus true, that is fiercely protected. Accordingly, no group of professionals, such as doctors, governed by a particular paradigm, will change until it has, Kuhn writes, "re-evaluated traditional experimental procedures, altered its conception of entities with which it has long been familiar, and…shifted the network of theory through which it deals with the world" (7).

Nothing, therefore, occurs overnight. There is not simply one counter-discourse that replaces completely the discourse it has vanished. In truth, there is a blending of forces, of ideas, of events, of rhetoric that lead to change, one point of power at a time turned to a point of resistance. Patients, parents, even doctors have spoken out for some time against established protocols. I mentioned Diamond and others objecting to Money's ideas as early as the 1950s. Even earlier, in the 1930s, when sex conversion surgery came to be practiced more frequently, parents who did not agree with the surgery suggested by the doctors refused to grant permission for any sort of medical intervention (Fausto-Sterling 92). But such voices were few and far between, and certainly they lacked sufficient ethos to create any sort of groundswell of organized, widespread resistance. Before postmodernism, Dreger argues, the opinions of those on the margins, those not in positions of power, did not matter much in the grand scheme of things. Certainly, a more top-down approach to issues, or one in which an institution or paradigm established parameters for normalcy, was the accepted norm.

But the advent of postmodernity, which opened up the opportunity for more intersexed persons to know their identities, to speak from new subject positions, and to find an audience that would listen (*Hermaphrodites* 170), saw the beginnings of messy change. Although the idea that there is only one true story, that there is only one legitimate authority, has not completely disappeared, it has been modified within the postmodern period so that those living and speaking from marginal or hybrid positions can now be heard. And if any person can be said to live in and speak from a hybrid space, one lost in between, and thus often ignored, it is the intersex person.

Dreger lays out a number of reasons why the flux of ideas and thought within postmodernism has ushered in a new period in which intersex persons have the means to know themselves and others like them. For one, the intersex person's perspective, or the perspective of anyone other than a doctor, was all but eliminated in the late nineteenth century, and the hermaphrodite's voice literally disappeared (*Hermaphrodites* 167). Because of postmodernism, however, intersex persons, laypersons, and even experts with viewpoints alternative to the authoritative position on intersexuality, previously silenced or ignored, have now become more prevalent and more influential, at times competing with and thus effectively resisting what had become the sole narrative regarding intersex persons and their treatment. In fact, because postmodernism rejects the grand narrative and posits instead that there are many narratives, each with its own truths, the paradigm for treating intersex persons has been increasingly called into question. Narratives previously silenced tell of secrets, of forced sex conversions, and of botched surgeries (*Hermaphrodites* 171).

Moreover, postmodernism has allowed intersex activists to describe their bodies as "colonized" objects, and this, according to Dreger, enables them, armed with knowledge and a voice, "to resist and object" to the medical discourse that victimizes them (*Hermaphrodites* 171). Thus, they regard medical treatments performed on them, whether that is performing surgery without the consent of the patient or keeping the condition that spurred such treatment secret from him or her, as institutional attempts to project societal conceptions of gender and normality upon them that they should and must resist.

Even if the doctor's attempts to normalize the intersex person are done with the best of intentions, "the modernist conception of the active physician hero," Dreger writes, "has given way to postmodern challenges of the doctor-patient balance of power and to challenges to the 'doctor as savior' motif" (*Hermaphrodites* 172). Within this new dynamic, intersex persons are no longer simply acted upon, but are themselves actors as well, actors who see themselves and their treatment as products of social and cultural constructs. As Dreger writes, they see that "it is not their bodies that make their lives difficult, but the cultural demands forced upon their bodies" (*Hermaphrodites* 173).

## A Focused Look at the Change in Rhetoric From 2000 to 2006

If we brought our focus closer in, using Dreger's references to the influence of postmodernism as a useful critical backdrop, I think we would see very specific reasons for why there was such an obvious sea change between the rhetoric of 2000 and 2006. One reason was growing intersex "street" activism efforts, begun in the decade before the 2000 statement, which had already started to nurture debate. Learning by accident, personal investigation, or confessions from parents or doctors, individual intersex persons began to come to terms with their identities and, in the process, seek out communities of self-support. Offline, then, various intersex groups began to organize, following a model of social activism born in the 1960s in response to the Vietnam War or the call for improved civil rights for blacks, then later reinvigorated by GLB and disability advocates in the 1970s and 1980s (Preves, "Out of the O.R." 182). It was and still is activism intent on championing new definitions, new self and societal perceptions of what it means to be intersexed, and new methods of treatment in opposition to previous approaches that left many shamed, if not physically and psychically scarred.

One of the first intersex activist groups was the Turner's Syndrome Society, founded in Minneapolis in 1987. In the United Kingdom, an Androgen Insensitivity Syndrome (AIS) support group was founded in 1988, and a US version in 1995. An association for those (or the parents of those) with Klinefelter's Syndrome was started in 1989 (Preves, *Intersex* 92). The Intersex

Society of North America (ISNA), until its abrupt, relatively shocking decision in July of 2008 (at the time this book was being revised) to shut its doors in favor of a new name, Accord Alliance, and a somewhat different mission, was perhaps the most politically active, if not the most influential community. As of 2006, ISNA, which was founded in 1992 by Cheryl Chase, had more than 2,600 participating members and hundreds of financial donors (Preves, *Intersex* 93). Chase was initially raised as a boy, but then, at eighteen months of age, was converted, with her parents' consent, to female (Fausto-Sterling, 80). She developed along with other ISNA members an agenda, shown as follows, that in large measure can be seen adopted in the revised 2006 AAP statement:

- Intersexuality is primarily a problem of stigma and trauma, not gender.
- The child is the patient, not the parents.
- Professional mental health care is essential.
- Honest, complete disclosure is good medicine.
- All children should be assigned as male or female, without surgery. ("Agenda" 240)

Other intersex activists, often with the same motivating experiences of unwanted surgeries and hidden medical histories, have attempted to forward this agenda or one similar in an effort "to destigmatize," Preves writes, "intersexuality and to alter the medical treatment they receive" ("Out of the O.R." 179). For example, Hermaphroditic Education and Listening Post (HELP) was formed in 1998 by a mother, Helena Harmon-Smith, who became enraged after a doctor performed surgery on her intersex child even after she had refused consent to do so. In forming HELP, Harmon-Smith developed a list of Ten Commandments that HELP and its advocates argue all doctors should abide by when treating intersex children. They include admonitions not to withhold information or support from the family, and not to perform unnecessary surgery, especially in the first year (Fausto-Sterling 92).

Yet another influential intersex group, the ISGI, or Intersex Support Group International, begun in 1998, draws on biblical references. It cites passages from the Bible to argue that intersexed persons are also God's children and that, therefore, current medical treatment of them, which often attempts to draw on religion for support, is in fact un-Christian in its approach (Preves, *Intersex* 93).

In 1994 ISNA began a newsletter, *Hermaphrodites with Attitude*, and, not long after, its membership grew. Then in 1997 ISNA publicly picketed a meeting of the American Academy of Pediatrics as well as hospitals where intersex surgeries were performed. This action led to mainstream media exposure for ISNA and its efforts, and from that point on ISNA representatives began to receive requests to speak at professional meetings regarding intersexuality, pushing activist rhetoric that began to cause some doctors to rethink the treatment they had previously given intersex children (Kessler 78–9).

It is this increased mainstream media exposure of intersex issues, as told from or sympathetic to an intersex person's perspective, which also can be seen as another reason why rhetoric shifted from 2000 to 2006. How? When Paul du Gay first introduced the "circuit of culture" (3), he used the Sony Walkman to illustrate the idea that anything, such as a product like a Walkman, or a concept like intersexuality, "means nothing in itself" (17). "What makes the Sony Walkman," what makes anything and gives it meaning, like intersexuality, "is not only the 'work' which has gone into constructing it meaningfully," du Gay et al. writes, "but the social practices with which it has become associated" (17). So, really, an entire collection of social practices, contributed via the media, among other things, serves to signify intersexuality, and to make that signification an ongoing process. Through the media, for example, and its more attentive and sympathetic coverage of intersexuality, especially in the period between 2000 and 2006, how intersexuality was represented, produced, and consumed changed, and along with it the regulations for those identified as intersexual also changed. John Colapinto's impressive book, *As Nature Made Him: The Boy Who Was Raised as a Girl*, which appeared in 2000, brought new light on the sad story of David Reimer, a patient of John Money's who became the test case for his theories on gender that led to the treatment paradigm that we see reflected in the 2000 AAP statement. In October, 2001 PBS aired a documentary, *Sex Unknown*, which, although controversial among intersex activists because not any intersexed persons were actually interviewed in the special, did nevertheless add more attention to questions being raised in the popular media about treating intersexuality (ISNA, "PBS Airs"). Other news brought increasingly harsher criticism, if not doubtful questioning, about early surgical intervention and sexual conversion in the treatment of intersexuality, and even the acclaimed novel, *Middlesex*, signaled both in its production and popular acceptance that new voices on the subject of intersexuality were in the mainstream—voices that reflected changes as well as served to produce them. The fact that the *Oprah* television program in 2007 conducted an hour-long episode dedicated to intersex issues as told from the perspective of intersex persons, when Money fifteen years earlier was a guest on the show (*Priestly Sins*) to forward his controversial views, illustrates that the cultural production of intersexuality had been influenced.

In addition to mainstream media, more critical focus in the academy, even in disciplines beyond medicine, was paid from 2000 to 2006 to the scholarly assessment of the various cultural, political, legal, and medical issues regarding the treatment of intersexuality. For example, although Dreger's influential work on the history of treatment is published in 1998, other significant examples, just to name a few, appear starting in 2000, including those this book already has referenced, such as Fausto-Sterling's *Sexing the Body* (2000), Preves' *Intersex and Identity: The Contested Self* (2003), and Kessler's *Lessons from the Intersexed* (2003).

What also is not coincidental is that the change I note between the 2000 and 2006 AAP statements takes place when the Web's popularity exploded. In 2000 it may well be that the dot-com bubble burst, but Web 2.0 and online social networking, as seen in the popularity of Myspace and Facebook, have exploded since with, in fact, the greatest growth in available social networking sites occurring since 2003 (Boyd & Ellison). More people now go online not just to buy things, but also to learn, or to commune with others like them. For example, 800 million more users were using the Web in 2006 than 2000 ("Internet Growth"). They were visiting or maintaining more than 75 million Web sites, compared to fewer than 10 million in 2000 (Zakon). They were searching for results that could be found on nearly 20 billion pages ("WWW Size"), not the three billion available in 2000 ("Number of Pages"). In short, between 2000 and 2006, the Web became for even more people from all countries, socioeconomic statuses and backgrounds a much more relied-upon place to visit regularly, and in that visiting to bring with them ideas and to have ideas brought to them that reinforced many long held beliefs, but that also ushered in new challenges to such beliefs. In short, although it is easy to get caught up in the numbers, often when they do not tell the entire story, in this case what they do point to is that the Web by 2006 had expanded exponentially, meaning there were new consumers, new producers, and new ways for representing and regulating meanings that were not available in 2000, making the Web, as a result, highly influential, along with all the other key reasons cited, in generating productive rhetoric that helped to bring about revisions to the 2006 AAP statement on intersexuality.

This Web rhetoric may be confrontational, as is the case of the ISNA agenda, mentioned earlier, that is posted for easy viewing on its home page. But other online rhetoric may be less "in your face" but still effective. In other words, because intersex Web sites give, Preves asserts, "former patients and family members an opportunity to discuss intersex beyond the clinic" (*Intersex* 95), the clinic is no longer the only place where intersex bodies or their conditions are controlled. The stories of intersex persons, their conditions and their treatments, are no longer unknown, meaning that parents like Gayle, who said, "When we had our daughter we were really isolated and we only had the so-called experts to talk to," have the opportunity to tell their stories, to get and give advice, to feel a little less freakish, a little less alone. Through the Internet, intersex persons or their loved ones can come together to learn more about who they are, who is like them, and what they can do together. Through this active creation of virtual communities of support, therefore, not necessarily overt confrontation, identity and also activism and resistance are born.

## TELLING STORIES: ONLINE INTERSEX NARRATIVE

Despite its limitations, the Internet can and does serve as a virtual meeting ground where intersex people can construct and nurture identities, both as individual intersex persons and as members of a community of others. There they can participate in a virtual discourse that provides alternative viewpoints, ideas, and opinions on the established treatment of intersex children. Put simply, they can tell stories that are heard, that are responded to, and that produce new truths, often not heard before, that resonate beyond where they occur, no longer hidden but available now via a simple search, a suggestion from someone who tells someone else to go to a specific site, that there are answers there or others that will understand and offer support.

In the early evening of March 4, 2007, Anna, who had just joined the Bodies Like Ours (BLO) Discussion Forum, sent out what could only be described as an online distress call. She titled it "New here, and confused" and attached a sad face emoticon to it to illustrate her mental state. Even though she indicated she was only 16, Anna had already endured years of doubt and ineffective medical treatment that had led her to feel increasingly isolated and desperate:

> I've come to a point in my life where I'm just so confused about my identity; I don't know what to do…I just need to talk to someone about this.
>
> I've feared I might be intersexed since I was nine-years-old and started to virilize/masculinize. I'm supposed to be female, so this was rather traumatizing and alarming. Needless to say, being a nine-year-old "girl" with a mustache and abundant underarm hair was not pleasant. I was teased mercilessly; it was a very emotionally painful time for me. I remember hearing about intersex conditions and being terrified that that was what I was. During this time, I tried to hide my developing masculinity by assuming an overly "girly girl" demeanor. I took to removing my body hair, hoping that this would lead to acceptance. It didn't.

When I was around twelve, I had a run-in with a doctor that has puzzled me to this day. I went (supposedly) for some sort of routine physical checkup. While there my genitals were inspected. At the time I just thought this was a normal part of the physical, and thought nothing of it. After discovering more about intersex conditions, I'm not so sure. I suppose I should have asked some of my more normal friends if they had to undergo their genitals being inspected too, but I was too embarrassed to bring the topic up. My sense of shame was only intensified by the doctor's comments to me and my mom about my appearance.

Today, at sixteen years of age, I am more confused than ever. What little feminization I had a few years ago has mostly disappeared. I look like a teenage boy, so much in fact that I am often mistaken for one. I'm the average height of a boy my age, have the same amount of body and facial hair, the same lanky body-build, an Adam's apple, a deep voice, masculine facial features…And I feel masculine. Basically I'm just your average-looking fourteen-year-old boy, except that I'm legally a sixteen-year-old girl. I actually thought I might be a transgender or a transsexual for a while, but I don't really fit in with them. Most of them have normal bodies, after all…I can wear a shirt without a bra, and my breasts are hardly visible. I don't look like a female, don't feel like one…What am I?

At this point in my life, I've given up trying to be normal. I am a freak, just like everyone has been telling me my whole life. I wear my hair short, dress in unisex and guy's clothes, and basically live my life as a male. People are afraid of me when they find out I'm legally female because I look so much like a guy, or at the very least androgynous. Often people express confusion over my gender; this has been happening basically my whole life. I would never wish this kind of life on my worst enemy. Some times I just want to die because I'm so confused, and it feels like I'll never resolve any of these problems.

> I've accepted my weird body, but I just want to know what's going on inside of me, so I can have some closure. I feel like my gender issues (both physical and mental) are consuming my life, and I just want to be able to find out what has caused this, so I can move on...

Less than two days later, the responses to her story came pouring in. One of those was from Andrea:

> You are wise to know you need support.
> You are smart to know where you can get it.
> You have guts to follow through and go get what you need.
> Clearly you are not someone who needs to apologize.
> I understand why you did though.
>
> It's hard dealing with teasing, from tools who tease you because they are also insecure and freaking out over their own developing sexuality, and the things said about you by insensitive adults, like some doctors and parents in your presence as if you were too dumb to notice or didn't rate more consideration...well that doesn't help either.
>
> Stuff like that would affect anyone's confidence! So you are not "defective" or something, and in fact far from it!
>
> Really I can't help but notice, that despite everything, you came here and spoke honestly from your heart. You have to be really something. Anyone would be very lucky to have you as a friend in real life. I envy them.
>
> I can easily relate to what you are going through, and there are many others who can also.
> You are not alone. ←- NOTE

Such is the power of the story for online intersex activism, and it illustrates how it contributes to the movement's alternative rhetoric. Anna didn't feel she had another place to turn to for support, but an intersex virtual community was there, and, through this alternative media, she was able to tell her story, and another like her was able to listen and respond, and others beyond that exchange, inside and outside of the community, intersex or not, were able to view it and be affected by it.

Being able to tell a story is a powerful act, especially when it could not be told before. Stories by many different names are the means by which we amuse, we express, we bond, we lead, and even by which we challenge other stories, other truths, with our own version of events, our own stories. Sharon Kilty, in fact, indicates that "stories are therapeutic for tellers who have a real opportunity to be heard and to hear themselves. As they tell and retell their story they can unravel the truth of their own experience… and begin to adjust to the person they have become."

Conversely, not being able to tell a story, or not being able to tell one that can be heard by others, means more than just not being able to express to others one's feelings. Because stories help to shape our perceptions of who we are and of what we believe to be right or true, those persons who cannot tell their stories lack power to influence beliefs, to challenge and shape the truth. In fact, if we agree with Michel Foucault that "critique is the movement by which the subject gives himself the right to question truth or its effects of power and question power or its discourse of truth" ("Critique" 194), then those who lack the power to tell stories, and to have them heard, lack, therefore, the power to critique, or to engage in the kind of "involuntary insubordination" that helps to contest and shape what all of us know to be right, or normal, or true.

Online, in a community (albeit virtual) amongst others with shared experiences, Anna can tell her story, and that story is heard. This not only impacts Anna, but it also resonates beyond because others can see Anna's story and be affected by it. Others in her situation may now not feel alone. Others, such as parents of children diagnosed with a particular intersex condition, might read Anna's story, or someone else's, or specific alternative recommendations written for them, not by doctors but intersexed persons, and these parents' views on what is the right way to treat medically or raise long-term an intersexed child might be altered. In the simple act of telling and sharing stories previously not heard, alternative truths are generated that counter the established truths that perpetuate the treatment of intersexed persons. Anne Hunsaker Hawkins calls such storytelling *pathography*. "Pathography," she writes, "returns the voice of the patient to the world of medicine, a world where that voice is too rarely heard, and it does so in such a way as to assert the phenomenological, the subjective, and the experimental side of illness" (12). Narrative pathography, therefore, enabled online, "restores," according to Hawkins, "the person ignored or canceled out in the medical enterprise, and it places that person at the very center" (12). "It seems that a common intersexual problem," according to Herman Stark, "is grappling with the haunting sense that one is not living as a true human self—or not living as the self truly is" (271). Stark argues that, within intersex narratives, "[O]ne can find...a call from intersexuals to let them suffer the truth of who they are." In other words, their stories online signal rhetorically that "life is one's own project, one's own projection." Therefore, Stark continues, "[A]uthentic existence means, in part, living out our own life, not the life others have foisted upon us" (277).

Such narratives not only allow intersex persons to tell their own truths, but they also lay the groundwork for effective rhetoric because the truths they tell often indict established truth. For Foucault, critique is "the art of not being governed" (193), and stories told by and to intersexed persons online represent, by what they say, who listens, and where they are said, significant instances of virtual critique and, in the process, resistance to the rules, the norms imposed upon us, our bodies and our identities. What Hall, in "Local and the Global," says of the identity struggle for those on the margins in Post-War Great Britain, such as immigrants from the Commonwealth, can be applied to the efforts of intersex persons:

> The emergence of new subjects, new genders, new ethnicities, new regions, new communities, hitherto excluded from the major forms of cultural representation, unable to locate themselves except as de-centered or subaltern, have acquired through struggle, sometimes in very marginalized ways, the means to speak for themselves for the first time. And the discourses of power in our society, the discourse of the dominant regimes, have been certainly threatened by this de-centered, cultural empowerment of the marginal and the local. (34)

First through individual acts, then collective street activism, and now through the Web, the truth about intersexuality has been critiqued. Productive rhetoric, through each story, each press release, each alternative view raised, has been engendered, and, in doing so, that which is supposed to be inconceivable—that there can be another way to treat intersexuality, that immediate medical treatment isn't always best, that living with ambiguous genitalia isn't viable—is, Butler writes, "conceived again and again" (*Undoing Gender* 64). In fact, intersex Web sites show to those who go there for support and those who perhaps go there just to lurk—that is, they show to all of us—that, quoting Butler, "there are humans," normal, functioning people, "who live and breathe in the interstices" (*Undoing Gender* 65) between what society perceives to be the "normal" man and woman. As a result, this virtual discourse, because it calls into question or attacks the very terms, concepts, or words that represent the so-called truths of established discourse, serves as a means to counter traditional medical discourse from within it, not outside of it, helping in the process to shift the paradigm governing intersexuality.

The productive nature of these virtual communities is the focus of the next chapter.

# CHAPTER 3

# VIRTUAL INTERSEX NEIGHBORHOODS

Katie

As far as the internet is concerned, at least for me, I originally found trying to find information on intersexed to be extremely hard to do. Back in 93, after finally being told i was IS [intersexed], true hermaphrodite, leaving the military a broken dispirited soul. It, the internet, was the only source of information, i had easy access too. When i questioned what i was, and found myself falling into despair, i found that at least i could find some medical journal, edited report about people like me.

It took me years to find a support sight [*sic*]. I believe CAHourstories.org was the first i found, that led to Locahsupport.org(this might be wrong, its [*sic*] been a few years), then Caresfoundation which led to hear [*sic*], so from 93 tell [*sic*]

2002, i only had found sites like Webmd, or Medical encyclopedias. And a lot of those only gave perspectives of how we are to be treated as an intersexed patient.

When i finally found other people that posted their own life stories did i actually find some form of acceptance. I finally found people who could understand me. That is something i needed, as i didn't find it with family.

At least i didn't know i had it, acceptance until a serious of overdoses. That is when i started getting answers, a few from parents, then a whole lot more from an aunt and uncle, they actually have provided me more knowledge about all that was done to me. My parents have sense refused to talk to me about what i am. I do know they love me, just wish they were more open with me.

The stability though that i have now, has a lot to do with knowledge i have gained from following other peoples posts. Here there is a large representation of many different conditions. People who are looking for answers, people who know what i don't, people who understand and have done marvels for my self confidence. Without this sight [*sic*], i'd still have been stuck in serious bouts of depression, if not worse (dead).

## DEFINING VIRTUAL COMMUNITY

Katie, along with a number of intersexed persons, their loved ones, or advocates, participate in BLO, a virtual intersex community and discussion forum. When Howard Rheingold first used the term virtual community in 1993 he attempted to explain the sense of community he and others felt by participating in a computer conferencing system called the WELL (Whole Earth 'Lectronic Link). Rheingold defined the WELL and other virtual communities like it "as social aggregations that emerge from the Net when enough people carry on those public discussions long enough, with sufficient human feeling, to form webs of personal relationships in cyberspace" (5). Catherine Ridings, David Gefen, and Bay Arinze avoid characterizing feelings generated when they define virtual communities as "groups of people with common interests and practices that communicate regularly and for some duration in an organized way over the Internet through a common location or mechanism" (273).

The virtual community offers legitimate opportunities to engage in support, in community building, and in activism. Sandor Vegh points out, "Since the traditional information channels may be well controlled by those whose interest is counter to that of activists, the Internet may serve as an alternative news and information source" (72). Even though a physical locale may not exist, participants in the virtual community "self identify with its content," Stuart Glogoff writes, "and seek out some degree of membership." What if there are just a few who are most active? It is the nature of every community that there are smaller, more active groups within each one, so that if, for example, a community has an association, then there are only an active few that work far more than most everyone else in the community to carry out activities that build its identity. Barry Wellman and Milena Gulia argue that most relationships people have in their lives are weak or lack face-to-face intimacy (13). Furthermore, weak ties according to Wellman and Gulia are not insignificant. In fact, because the Internet lacks social cues, such as gender, race, or class, that often hinder a sense of shared community when people with weak ties interact in-person, any lack of intimacy not shared by those corresponding virtually is made less important. If anything, the lack of intimacy cannot be a barrier online when sometimes it is offline. A person's age, wealth, and so on, therefore, are not valued. What might stand in the way of two people forging a tie or relationship in the physical world is done away with in a virtual environment. Online it is what people have to offer the community that is valued most (8) and that also identifies them. So just as online there are lurkers, there are also offline members of communities who live on the street but don't put up Christmas decorations, who turn their lights off when the local kids come by to trick or treat on Halloween. These people don't actively engage with their neighbors—they just live there. Yet they are there, they listen, they absorb, and eventually some even become more active. And when they become comfortable enough "to share intimate details of their lives with a virtual stranger," Glogoff continues, "it affirms that an implicit context of trust has been established." It is this trust as well as the kinship it produces that serves to bind together the virtual community.

Perhaps one never knows the identity of those other people one communicates with in the virtual community. Do we really know that Anna is 16, that Katie is telling the truth about her experiences, that either one is really who they say they are? Of course, no such guarantees exist online, but such anonymity can also be regarded as important to fostering the intersex virtual community or other similar communities that offer support to those who have been abused, stigmatized, or isolated by their bodies or their identities. Poked and prodded, subjected to medical examinations and procedures without adequate explanation for why, "[T]he Internet," Preves writes, "has been especially important for people to seek out others in a relatively anonymous forum" ("Out of the O.R." 188). As Batya Friedman et al. notes, "[O]nline anonymity represents an opportunity for important interpersonal relations. For example, a gay teenager in an intolerant family or community might rely on the anonymous characteristics of the Web to find liked-minded peers online."

Just one discussion from the BLO forum ideally illustrates the importance of anonymity for those who have been scarred so much that they want to tell their stories but still feel the need for some degree of protection. Dorie writes,

> A certain degree of privacy (not using my legal name) allows me to speak more freely and the Internet provides a forum where I can do that. After a rough childhood and adolescence I was able to put things right and go on with a normal life in my early 20's (well over 30 years ago). A few of my closest friends know I was IS, though one of my close male friends just can't get his head around the concept and still thinks I was/am a transsexual.
>
> In my 50's now, I appreciate my anonymity and would just as soon not deal with an ignorant public who are prone to conclusion jumping! So I keep my distant past and my "medical oddities" to myself.

Following less than an hour later, Penny posted concurring thoughts about the importance of anonymity:

> I stay relatively anonymous…If you walked up to someone, a stranger, and told them you were xxy most people would not have a clue what you were talking about and as you explain it to them they probably would not really understand still.
>
> Now if I walked up to someone and stated that I was a hermaphrodite well now, not much explaining there. We are famous for porn stars, sexual deviants, freaks of nature, abomination of God, carnival freak shows, and my favorite "chicks with dicks."
>
> I can't fight years of that crap. I need to work and I don't want my house burnt to the ground. So I just keep that to my self and life goes on.

It may be that those online communities that share kinship only because of a single, possibly fleeting, fad, or interest might be considered too loosely formed to be legitimate, and it is perhaps the case too that those participating in such communities see them as diversions to their offline lives. But such conditionals do not apply to the intersex virtual community. Even online participants they are marked by their conditions, their bodies, their treatments, and their stories. These things, such as Penny describes, cannot be left at the door on the way to the computer. In fact, the virtual community they go to is not a diversion, and it is not a substitute for something better that can be had offline. It is, for so many scarred by so much, one of the few places where they can feel protected enough to find solace, to get advice, to be with others who share with them many of the same things that once if not still continues to isolate them, physically and emotionally, offline. Listen to Dorie again as she writes of the value of an intersex virtual community for her:

> I am 57 and didn't know I was Intersexed until a few years ago when a whole bunch of "dots" suddenly connected and everything suddenly made sense. I was **INCREDIBLY ANGRY!!!** I was angry at the doctors who hid my "deformity" from my birth mother and mutilated me when I was an infant without anybody's permission. I was angry at my adopted Mom for trying to force me into a role for which I was totally unsuited and for preventing me from seeking medical help. I was angry with my adopted parents, the extended family and family friends who knew about my DSD and never told me even when I was going nuts trying to understand why I didn't "fit."
>
> I had never met any IS people (that I know of) and my Internet search took me to ISNA and that led me here. This was (and is) the only forum I have to meet others who understand and whom I understand and that helped me deal with my anger.

Such stories are not meant to idealize virtual communities to such an extent that we believe they are better than anything that can be done offline. But there is anonymity online, and there is also a nonjudgmental community of people just like them who will not examine intersexed persons medically as though they were a subject for study, who will not be shocked by what they have to say, who will listen and support them. Where is the activism in that? It may not be that the people who go to such intersex virtual communities see themselves as "part of a social movement," Patricia Radin writes, "[but] they nevertheless are using the global communication grid of the Web to connect with like-minded people" (592). In this connecting, in the exchange and sharing of information beyond the control of the clinic, intersexed persons and their advocates become empowered. As this mutual support is fostered, Radin argues "it is only natural that some will identify targets for social action" (592), such as poor medical treatment. Via support for each other, therefore, for building and sustaining community, virtual intersex activism is born.

## Porter's 5 Ps of Virtual Community

To try to understand the nature and shape of the intersex virtual communities that can produce such activism, I rely on what Constance Elise Porter has proposed as the "Five Ps of Virtual Communities":

1. Purpose (Content of the community)
2. Place (Extent to which technology mediates interaction in the community)
3. Platform (Design of the interaction)
4. Population (Participants, their patterns of interaction)
5. Profit Model (Return on interaction)

Now, at first glance, not all of Porter's Ps may seem applicable. An emphasis upon profit, for example, may not have much value in measuring the dimensions of a virtual intersex community. But Porter's essay provides a useful means of analysis for my work. For one, it enables me to define virtual communities according to certain key attributes. In addition, because the nature of each community's attributes can be different, I can compare one community to another—either one virtual intersex community to another, or intersex communities to other types of virtual communities.

For Porter, the spaces where virtual communities reside are places, formed through bonds and communal norms, that have boundaries just like homes have walls or neighborhoods have borders. The quality of those boundaries vary with each virtual place as they do with each neighborhood, and so it is possible, applying Porter's Five P's, to declare that intersex virtual communities are not just different, but that some are also better at providing a sense of place and, thus, community for its participants. For example, as important as the particular virtual community's purpose is in contributing to what makes it different, the design of it, not only how it looks but the means of interaction it provides its participants, or its platform (Porter's 3rd P), is even more crucial. A community, in other words, that offers a discussion forum, as opposed to one that does not, is clearly one that is more capable of offering support and of engaging in activism via it because participants in it have more opportunities to engage with each other.

What also cannot be discounted is that such a community is more "state-of-the-art," more professional, and thus more appealing to those interested in joining and more credible to those, outside of the community, who might view it to use the information it provides. For example, a comparison of a virtual community with a technically sophisticated platform, such as the one maintained by BLO, with the platform of the Hermaphrodite

**FIGURE 1.** BLO Web site.

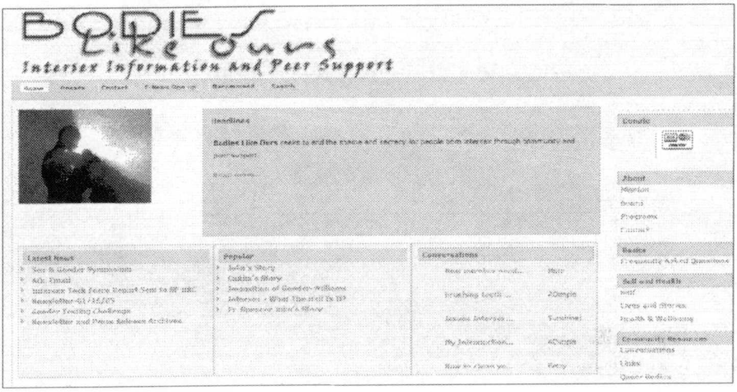


Education and Listening Post (HELP), illustrates this. The BLO site, which can be seen in figure 1, has a polished feel, uses cascading style sheets to lay out its information in a clean, appealing manner, and even includes a photo graphic and layered color scheme. It has hierarchical navigation running across the top and down the right side of the user screen. This offers users continually deeper layers of information about a wide variety of topics they can access to any depth or in any order they want. In addition, it uses software to allow members to communicate with each other via chat rooms or discussion boards. The most popular or recent postings (called "conversations") from members as well as articles relevant to intersex persons are also automatically listed on the site's main page using programming language that pulls this information from a database working behind the scenes.

In short, the BLO virtual community appears different than HELP's (shown in figure 2), which, although offering information and some interactivity, clearly lacks the same sort of

**FIGURE 2.** HELP Web site.

HERMAPHRODITE EDUCATION
and LISTENING POST

**What is HELP**

HELP is a support group for parents, family, friends and persons affected by Sex Differentiation Disorders. It is a source of medical information, literature and personal experiences. The goal is to connect persons affected by these disorders, to educate the public and medical profession, and to offer emotional support and understanding to those facing the decisions involved with these disorders.

- Who founded **HELP**
- Where is **HELP**
- HOW does **HELP** help?
- A list of Support Groups

advanced technology, the same depth of features, and the same quality design.

In terms of platform, therefore, although BLO and HELP are both intersex virtual communities, they are not only clearly different from each other, but the BLO community has a more professional look, and beyond this design interface it provides an active discussion forum that allows for better participant interaction and, therefore, a better sense of community.

The nature of this community, or what Porter's 4th attribute characterizes as its population or pattern of interaction, is determined by the ties that bind its members, whether they are weak or strong, and also upon the community's group structure, whether it is a small group or a larger network of individuals. An intersex Web site offering discussion boards or chat rooms, as does BLO, creates a stronger social connection, one that encourages more open, if not real-time, interaction, than does a site that only allows e-mail to be sent to a site administrator who responds to the e-mail but not publicly.

In addition to the type of interaction, this Population attribute also seeks to describe who is in the community. How many are there? Do the same people come to the site? How often do they come? Are the people meeting in this virtual place also participants in other communities and thus part of a larger network of virtual places, or do they constitute an isolated group that just meets there?

These are difficult questions to answer without proper tracking methods in place. Because of this, the more technologically sophisticated virtual communities use Web tracking software, along with direct surveying of their participants and members via e-mail and other methods, to determine who they are and what interests them. For example, the BLO cofounder and Web site administrator, Betsy Driver, was able to answer quickly (when asked in an e-mail interview) that her Web site had 197,492 unique visitor sessions (each session representing a different user on the site) in one year, 356 of these visitors were registered members who actively participated in site-sponsored chat rooms and discussion forums, and 496 more were signed up for e-mail newsletters and other distributions. If a sense of community is built upon not just a shared interest but also upon member awareness of if not interaction with each other, then intersex virtual communities such as BLO, armed with the ability to track and record who their members are, have a grasp of their population that rivals that of physical communities. Even those intersex communities without such methods of measurement can almost take for granted their membership in ways that other virtual communities cannot. Whereas other communities might have a lot of visitors who come and go, who might be curious for awhile and then leave, who might be motivated but then move on to a different phase in their lives, the life experience of the intersex person binds them to the virtual intersex community in ways that are not easily forgotten or cast off.

Some virtual communities, constituted by the interaction of its members, produce revenue or are formed for the purpose of generating profit, which is the 5th and final P for Porter. Again, however, such a notion of profit doesn't really apply to the virtual intersex community. But what if for the purposes of my study here we used Porter's Profit attribute to refer to other sorts of nonfinancial profit that can be gained, both by members individually as well as the virtual community itself? Through their interactions, members of intersex virtual communities produce emotional support, kinship, and effective activism that is profitable for them personally, for the virtual communities in which they interact, and for the intersex movement as a whole. Profit, therefore, can also be measured in terms of a virtual community's rhetorical value. Put differently, can one intersex virtual community be seen as more valuable than another because it constitutes a virtual place more capable of transformative, meaningful counter-discourse than another?

The answer must be yes, and this fact must further serve to define intersex virtual communities as different not just from other sorts of virtual communities that do not seek such change, but also from other intersex virtual communities that lack the potential to seek the change they desire. As Porter's 5 Ps allow us to see when applied, BLO possesses its own kind of population, and its potential for producing profit, for reaching out to its population, for offering a stronger sense of place and purpose, for, in other words, creating a more clearly defined virtual community is arguably much better than what HELP can do, simply because its design is lacking, it does not have the same interactive capability (such as an actively managed discussion forum), and even its focus is far more narrow.

Extending Porter, can it be said that a site like BLO's, with better design, better information, better organization, and better technology that allows for better, if not more, member communication, is one that has the potential for better profit, or for producing legitimate counter-discourse? As important as Porter's analysis is for my efforts, it doesn't answer this question and I cannot fairly attempt to make it do so. So another critical approach is necessary, one that analyzes not just what virtual communities do to be what they are, but also what forces outside the communities shape them and, in turn, are shaped by them through the counter-discourse they produce.

## Appadurai's Virtual Neighborhood—Product and Producer

Such an approach is made possible by relying on the ideas of Arjun Appadurai. For Appadurai, the world cannot be simply explained in binary terms where clear dividing lines separate nations, economies, and cultures from each other, the haves from the have-nots, the *One* from the *Other.* Because of deterritorialization, or the uprooting of people, of economies, and of cultures from once seemingly fixed positions via significant changes brought on by enhanced technologies, new modes of production, and altered political systems, the world's cultural system is more rhizomic in nature. In other words, what we think and believe is not isolated to the place where we reside. One culture is not cut off from another. Global flows, full of irony and resistance according to Appadurai (29), have unfixed culture from previous moorings, and as people, ideas, images, and money move freely around the world, so does culture, resulting in a web-like system where people and the local discourses in which they live and that shape what they believe are products as well as producers. They shape themselves, but are also shaped by others outside them whom they, in turn, shape. So the flow of people, of information, the widespread distribution of media, and the greater distribution of economies on a more global scale have brought about a world where the imagination and thus identity have become deterritorialized. The result is that what we think and who we imagine ourselves to be is influenced by local and global contexts, and these contexts or cultural flows overlap, subvert, resist, and influence each other.

To better explain the cultural landscape that the individual resides in, shapes, and is shaped by, Appadurai uses the terms *ethnoscape*, *financescape*, *mediascape*, *ideoscape*, and *technoscape.* A technoscape, most relevant here, refers to the flow of people who are influenced by technology. Because of particular technological tools, such as the Internet, people from different places interact with each other, and in this interaction they exchange information and values. The communications that people share via technology give shape to a landscape where people from different places and, thus, different cultures come together and create what Appadurai refers to as "imagined worlds…that are constituted by the historically situated imaginations of persons and groups spread around the world" (33). In other words, it is within one of these particular landscapes, not an isolated locale cut off from all others, that individuals imagine, think, and live.

In the case of a technoscape, those people who may otherwise be separate from each other's physical space, such as intersex persons, use a tool of technology, like the Internet, to create virtual communities where previous ideas, once tied down or so localized that they are marginalized, are contributed for discussion to be rejected or to become influential. Because others can, like the general public or media, use this same tool to participate in this discussion, they are also part of the same technoscape, capable of contributing to it and being influenced by it. So concepts fixed to a previously unquestioned discourse are put in play and debated. Notions of objective, universal truths, such as those regarding gender that greatly influence Money's medical paradigm for treating intersexuals, are examined from unlimited perspectives, and their cultural and historical origins are brought to light, questioned, and even countered.

Appadurai's critical perspective allows us to see that we live in fluid, irregular places, places that are shaped by changing ideas, persons, and values inside and outside of them. As a result, our places are "deeply perspectival constructs, influenced by the historical, linguistic, and political situatedness of different sorts of actors" (31). Consequently, our sense of place depends upon our imagination of it, and our imagination of it is a construct that is shaped by a variety of agencies, not limited by a fixed location, but all the product of a greater global cultural landscape.

What comes from all this is the idea that how we imagine ourselves is not about fantasy or escape. Imagination, whether it be how the individual or what Appadurai calls the "neighborhood" imagines itself, is an "organized field of social practices…and a form of negotiation between sites of agency (individuals) and globally defined fields of possibility" (31). Imagined worlds to which we contribute and belong are not isolated but are, in fact, interconnected, shaped, and influenced by other worlds they also shape. Therefore, when intersex activists produce an imagined world of their own in reaction to their treatment, that world, built on the social practices of a shared imagination, can "contest and sometimes even subvert the imagined worlds of the official mind…that surrounds them" (Appadurai 33).

From this critical perspective that Appadurai lends us, we can see how and also why intersex virtual communities are capable of producing counter-discourse. Technology allows intersex individuals to create online imagined worlds, and when they imagine themselves differently than does the imagination of official, established discourse, then their shared imagination serves as ironic disjuncture, resisting if not breaking up the practices and protocols that have negatively contextualized intersexed persons.

The possibility and thus power behind counter-discourse for Appadurai comes in the imagined worlds or "virtual neighborhoods" that intersex activists create. "No longer bounded by territory" (195), these virtual neighborhoods are electronic versions of the physically tangible neighborhood that Appadurai argues "is a context, or a set of contexts, within which meaningful social action can be both generated and interpreted" (184). So, when the Internet allows for intersex persons to communicate online with each other, they contribute individual imaginations to the creation of a shared communal imagination.

Through this, imagining intersex persons create a context or neighborhood that provides meanings, values, and practices that help participants of the neighborhood to see themselves as they want to be seen, to build bonds of kinship, and to participate in social projects, such as online discussions on relevant topics, that continually allow for the neighborhood to redefine its shape and purpose. As the intersex activists produce meanings for themselves, or "engage" in what Appadurai refers to as "the social activities of production, representation, and reproduction (as in the work of culture), they contribute…to the creation of contexts that might exceed the existing material and conceptual boundaries of the neighborhood" (185). Therefore, their very acts of imagination, which they engage in to produce themselves, also serve to produce meanings that, because of global cultural flows, go outside their neighborhood, impacting if not changing contexts or neighborhoods beyond their own.

## Making Noise—<br>How Intersex Virtual Communities Resist

How do we determine the quality or effectiveness of the rhetoric that intersex virtual neighborhoods produce? Frankly, it is hard if not impossible to measure quantitatively the possible positive impact intersex virtual communities have, by themselves, in producing counter-discourse. How many members are required to participate in online discussion groups for the Web site to be considered successful? How much money does it need to raise? How many medical practitioners have to change their methods of treatment, adopting alternatives proposed by intersex activists?

According to Amy Koerber, trying to determine answers to such questions is the wrong way to go about assessing the effectiveness of the counter-discourse produced by virtual communities. Koerber, in looking at Feminist Mothering Web sites, argues that these sites are effective in terms of resistance, not because they are politically meaningful in a traditional sense where obvious gains are made and documented, but because they make "noise" (219). In other words, Koerber argues that "resistance must be conceived as a process of making noise that undermines dominant discourse by getting inside them" (225). Feminist Mothering virtual communities, and others that make noise like them, offer discourse "that can be heard by other like-minded individuals, and, in doing so, they can produce noise, or ideological disruption, in the mainstream discourse that dictates reality in both on and offline worlds" (236). Therefore, resistance to traditional discourse can occur, according to Koerber, "when something said in cyberspace unsettles the common sense of the offline world, when the sense-making scheme of the two worlds collide...and the resulting noise is neither tuned out nor transformed into music, but...is listened to" (238).

Lorraine Code's concept of discursive space enables a critique of the intersex Web sites in which the very act of being online and talking differently or against mandated discourse constitutes resistance. By applying Code's ideas then, it can be argued that intersex Web sites are virtual "gaps" in the discourse, "not fanciful or fixed locations," but, instead, discursive spaces where intersex people and their advocates can have a reasonable "expectation of being heard, understood, taken seriously," simply because they are in a place, albeit virtual, where others like them can go, support, and resist (x).

Barbara Biesecker, following generally along the same lines as those laid out by Koerber and Code, relies on Foucault's understanding of power to assert that those seeking to resist power, such as intersex people and their advocates, must be productively resistant. To be resistant, what they do, or their discursive practices, "must not make sense within the available lines of intelligibility or discernment. That is, they do not signify" (357). Accordingly, Biesecker argues that alternative power or counter-discourse can only be produced through "resistant practices...that defy translation, throw sense off track, and, thus short-circuit the system through which sense is made" (357).

When the Intersex Society of North America's (ISNA) Web site posts a list of standards for intersex medical treatment that counters point-by-point the standards of established medical protocol, ISNA seeks to undermine, to resist, to defy if not replace what is taken for granted or understood with new translations. In doing so, it engages in a counter-discourse as a means of asserting power, from a virtual position but one with community support, which resists the power-knowledge of established discourse.

Such discursive "sense-unmaking" is at work also on the Web site of the Intersex Support Group International (ISGI). The ISGI site offers a fascinating "Apologetic" that is meant to counter the norm often forwarded in discourse that suggests intersex people are physically as well as morally abnormal, especially those who wish to continue through life in-between, not choosing to be either male or female. ISGI's online Apologetic argues that intersex people, rather than an abomination before God, are, in actuality, "children of God." The ISGI Apologetic, in fact, usurps discourse by quoting the bible for support, pointing out that in Matthew 19:12 Jesus said, "Some are born eunuchs." Therefore, if Jesus said there were some born eunuchs, then it cannot be said that eunuchs or intersex people go against God or nature.

ISGI argues that medical treatment, rather than giving a normal identity to intersex children, has robbed them of their true "life identities." The Church, according to ISGI, has also participated in this oppression. And it is true that God has certainly been used, recently and in the past, to buttress thinking on intersexuality, further cementing the notion that God made man and woman and nothing in between. Discursively, ISGI uses the same bible used against intersex people to show that alternative scripture exists. Again, applying Foucault discursively, we see ISGI as well as ISNA messing with the signifiers, disturbing previously held notions of meaning and thus making noise.

How do we determine the nature of this noise? How can we describe the productivity or rhetoric intersex virtual neighborhoods produce? The construction of rhetoric is like any kind of construction. It can be said to be constituted of certain parts and materials, to have a certain architectural design. Rhetorical acts, what Koerber calls noise, such as telling stories or offering advice, can be called the materials, the metaphorical nails, wood, and concrete that construct and maintain productive rhetoric, enabling the logos, pathos, and ethos necessary for rhetoric, including the rhetoric of intersex Web sites, to be effective and powerful.

But how does this construction occur, and what does it look like, what does it mean, upon completion? How can we describe its design? We cannot just count all the nails and make valid conclusions about design from that, just as we cannot count the stories to understand the rhetoric of intersex Web sites. Such a focus is too limiting because it does not adequately illuminate the entire process. As du Gay et al. notes, "[T]o understand production processes," whether that involves the making of technology or even the use of it in reconceptualizing something like intersexuality, "we need to refer to more than…the definition of specific activities" (43), like hit counts or member activity. Rather, "What we need to try to understand," du Gay et al. continues, "are the distinctive practices used in the production…and the way that such widespread practices are represented in terms of specific values, beliefs and patterns of working" (43). In other words, to go beyond just saying that intersex Web sites un-make our sense of things, to go beyond saying that certain sites have so many unique visitor sessions per year and that must mean they are active, we need to address what du Gay et al. characterizes as their "culture of production" (43).

Although there are probably many ways intersex Web sites and their participants contribute to new representations of intersexuality, for this book I focus on three: bonding, disseminating, and confronting. For intersex Web sites to be culturally productive communities, they must create a sense of belonging, a feeling among their participants that common values, if not viewpoints, are shared there, along with common experiences, as told often through stories, which bond them together. For these communities to contribute to how meaning is represented in culture, they must also be able to produce and disseminate alternative meanings. Finally, the information they disseminate must be confrontational so that they can resignify previous conceptions of intersexuality.

According to du Gay et al., to understand the Walkman as what it is, we need to understand more than its hardware. We must know the culture that produced it. Not just the internal organizational culture of the organization, Sony, that invented and manufactured it, but the larger culture beyond it that perceived it and through consumption of it contributed also to the regulation, production, and identification of it (43). This is also true for intersexuality, just as it is true for any intersex Web site, or for that matter any concept. To get at how anything in culture is produced, we must look at its process. So to understand intersex virtual neighborhoods, we focus on bonding, disseminating, and confronting because these processes, when understood, allow us to describe, as du Gay et al. describes the Walkman's production, how intersex Web sites establish cultures for themselves that provide them a shared platform through which and by which they produce new representations of intersexuality that are manufactured for consumption by a wider culture in ways not possible before, in ways that have the potential to lead to change in that culture. As Hall writes, "[T]he most profound cultural revolution has come about as a consequence of the margins coming into representation" ("The Local and the Global" 34). An examination, therefore, of bonding, disseminating, and confronting in the next three chapters, when viewed as culturally productive practices, offers concrete insights into the productive noise intersex virtual neighborhoods make from the margins.

# Chapter 4

# Bonding

Shani Orgad writes, "Internet spaces offer unique opportunities for diasporic members to facilitate, maintain and intensify their sense of community by reproducing and maintaining cultural elements of the homeland" (880). Now, intersex persons don't necessarily have a homeland, but the Internet works in much the same way for them in fostering their sense of community. Wellman and Gulia note, "[T]he ties that people develop and maintain in cyberspace are much like most of their 'real life' community ties" (16). Such ties are certainly produced by various means on intersex Web sites. Whether it is done through supporting the same agenda, offering advice, or sharing stories about the lived experiences of their bodies, intersex persons build associations or relationships with each other that serve as the foundation of culturally productive intersex virtual neighborhoods.

Without the ability to form communal bonds in virtual neighborhoods, many intersex persons would be too stigmatized or too isolated by their geographic location and treatment to come forward and interact with each other. The Internet, in other words, offers the chance for many intersex persons to bond with others like them when otherwise they would never have that opportunity. Lynnell Stephani Long, an intersex activist and educator, speaks to the effectiveness of the Internet as a means of fostering community:

> I think the Internet has been very helpful in reaching out to people who are still dealing with personal issues which may prevent them from seeking out someone in real time. Not all adults that are Intersex, and seeking information about their condition, have overcome the fear and secrecy of being Intersex. Thanks to the Internet, people don't have to leave the comforts of their own homes to gain vital information. ("Question")

Emi Koyoma, director of Intersex Initiative, points out, "[C]oncealment and isolation are core tactics of the traditional medical treatment of discourse" ("Question"). Very often, intersex persons are so cut off even from knowledge of their conditions that they do not know what they have, not to mention the fact that others also share those conditions. The Internet, simply by allowing intersex persons a place to bond with each other, undermines efforts to conceal the truth of intersexuality. "Without the Internet," Koyoma adds, "it would have been impossible to connect these deeply concealed and isolated people."

## MISSION STATEMENTS AND AGENDAS

One way we know intersex Web sites are communities is the mission statement or agenda many display prominently. Although similar purposes are articulated offline as well, as the accomplishments of ISNA and other intersex groups demonstrate, a visible, shared purpose online, which mission statements and agendas reinforce, serves to create bonds among persons in a virtual neighborhood. Purpose, such as advocating against unwanted genital surgeries, not only provides a sense of continuity among intersex Web sites and those that meet there, but it also brings people together and keeps them active members and thus contributors to a virtual neighborhood.

Some organizations, such as ISNA, are more advocacy-oriented than others and their stated missions or purposes reflect this. ISNA asserts, for example, that its mission is "devoted to systemic change to end shame, secrecy, and unwanted genital surgeries for people born with an anatomy that someone decided is not standard for male or female." Therefore, the overriding purpose at the heart of the ISNA Web site, one that binds its members together in a common cause, is the effort to once and for all do away with the old methods of treating intersex children. The BLO mission statement reflects a similar purpose: "We believe that through awareness, education, and peer support, our goals of eliminating isolation, shame and secrecy experienced by people born with Bodies Like Ours will be reached." Here, the use of first-person plural pronouns to convey that purpose is interesting, since it is not clear if "we" is meant to be exclusive or inclusive. But this vagueness may be purposeful, allowing those who are not, or who do not consider themselves to be intersexed, to consider themselves part of this virtual community. Koyoma argues as much when she writes,

> [M]ost people with intersex conditions view themselves having AIS or CAH or some other condition, and not necessarily "intersex," they would not join an "intersex" online communities. (In fact, this is the reason BLO started—it wanted to reach people who would never

> talk to ISNA because they don't consider themselves "intersex.") ("Question")

Purpose and the rhetoric used to state it help decide who considers themselves part of the community and who does not. Those sites with purposes more closely tied to a particular condition obviously construct boundaries that exclude some intersex persons. For example, The Androgen Insensitivity Syndrome Support Group (AISSG) Web site provides "information and support to young people, adults and families affected by complete and partial Androgen Insensitivity Syndrome (AIS)." By contrast, AAKSIS, or the American Association of Klinefelter Syndrome Information and Support, focuses just on Klinefelter syndrome (males are born with an extra X chromosome—XXY). The Turner Syndrome site is for those who have, or are family members of those with, an XO chromosomal makeup. Such statements ensure that those visiting these sites will understand the goals of each group, and the goals will serve to unite people in a common cause. Persons not committed to the stated goals of a particular site will go elsewhere. Just as communities include they also exclude, and having a purpose as voiced through mission statements or agendas allows visitors to determine whether they wish to be included or excluded.

### HYPERLINKING

Beyond the community each intersex Web site creates for its participants is the wider intersex virtual community, one that continually grows through hyperlinking. For example, most of the intersex Web sites examined in this chapter include links to each other, if not to other related intersex activist groups. The Intersex Initiative Web site breaks links down into specific

**FIGURE 3.** Debbie Hartman's personal Web site.

categories, including intersex activist sites like BLO or ISNA; condition-specific sites, such as AISSG; and even individual sites maintained by intersex activists. These individual sites not only offer accounts of personal experiences and stories that some visitors to them can identify with, but they also link back to the main intersex Web sites that link to them, so that visitors who come first to individual sites can also access the main group sites via hyperlinking.

One such personal site belongs to Debbie Hartman, an emeritus board member for ISNA. A screen shot of her site can be seen in figure 3.

Debbie's site tells just one story, a personal one about Debbie's daughter Kelli, who is an intersex child:

> Kelli has had 4 surgeries to try and make her genitalia appear cosmetically correct for a female. Kelli was subject to these early surgeries due to my uneducated nieve [*sic*] consent…I now know that each intersex person has their own right to choose if they want surgery or not.

> Not their parents, their doctors - no one! Kelli has both physical and emotional scars due to these suregeries [*sic*]. I now advocate for intersex people everywhere. We must help educate the world about intersex. Rid the shame and secrecy that has haunted these individuals for years.

In isolation, Debbie's site might not have much impact. It suffcrs from poor dcsign and cvcn spclling crrors. But if we look at Debbie's site as just one example of countless other sites created by or for intersex persons, all connected via hyperlinks with the Web sites of intersex activist groups, then we see a network of multiple, interconnected nodes that constitute not just a single intersex community focused on a specific condition, but rather a larger intersex neighborhood, one capable of greater rhetorical influence.

Resistance, again, is not radical or revolutionary. It is, as Foucault asserts, "transitory," occurring here and there, attacking and usurping traditional discourse not because of some master plan, but instead for personal or specific reasons. Such reasons include the desire on the part of intersex persons to let others

**FIGURE 4.** Navigation links from Klinefelter Syndrome Support Group home page.

| | |
|---|---|
| Support Group Information | National & International Conferences |
| Variant and Prenatal Diagnosis Information<br>(47XYY, 47XXX, 48,XXXY, 48,XXYY, 49,XXXXY & Prenatal diagnosis) | E-mail lists for support and information |
| Current Research Studies | Informative Resources |
| Klinefelter Syndrome Organizations | Klinefelter and Variant Websites |
| Online Pharmacies | Resourceful Genetic Websites |
| Klinefelter & Variant web stores | Klinefelter Syndrome Awareness |
| Symptoms & Educational Information | |
| Websites for non-Klinefelter links especially other genetic condition | |

know their stories, to find others like them, and also simply to engage simply in community-building online when it is almost impossible to do so in the physical, offline world.

When intersex sites hyperlink not just to other group and personal intersex sites, but also to those offering research, news, or support regarding intersex-related subjects, users gain a greater awareness of who they are and the community to which they belong, which also helps build a sense of community. The navigation links from the Klinefelter Syndrome Support Group (KSSG) home page (see figure 4) show this.

Most of the main links on KSSG home page actually direct users to pages that contain links organized by subject. Users can find links to research studies on Klinefelter's Syndrome, or links to sites containing information on other genetic conditions. There are also lists of links to support sites, like the list that can be found by clicking on "E-Mail lists for support and information."

**FIGURE 5.** List of support links on Klinefelter Syndrome Support Group Web site.

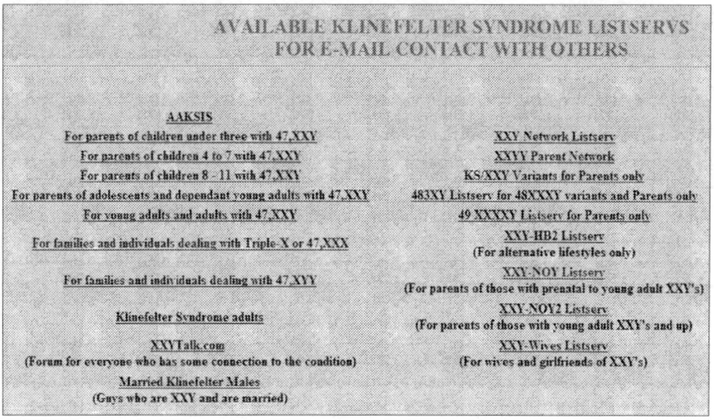
AVAILABLE KLINEFELTER SYNDROME LISTSERVS
FOR E-MAIL CONTACT WITH OTHERS

AAKSIS
For parents of children under three with 47,XXY
For parents of children 4 to 7 with 47,XXY
For parents of children 8 - 11 with 47,XXY
For parents of adolescents and dependant young adults with 47,XXY
For young adults and adults with 47,XXY
For families and individuals dealing with Triple-X or 47,XXX
For families and individuals dealing with 47,XYY
Klinefelter Syndrome adults
XXYTalk.com
(Forum for everyone who has some connection to the condition)
Married Klinefelter Males
(Guys who are XXY and are married)

XXY Network Listserv
XXYY Parent Network
KS/XXY Variants for Parents only
483XY Listserv for 48XXXY variants and Parents only
49 XXXXY Listserv for Parents only
XXY-HB2 Listserv
(For alternative lifestyles only)
XXY-NOY Listserv
(For parents of those with prenatal to young adult XXY's)
XXY-NOY2 Listserv
(For parents of those with young adult XXY's and up)
XXY-Wives Listserv
(For wives and girlfriends of XXY's)

Figure 5 displays the links found on that page, all of which offer additional support for visitors to the KSSG site.

As figure 5 shows, users can reach out to a variety of support groups, thereby finding one or more that meet their specific needs.

## Billboards, Discussion Forums, and Member-Only Access

To foster further a sense of community, some intersex Web sites offer features on their sites that resemble billboards found at community buildings. One example is the main discussion forum page for BLO. As figure 6 shows, there is something for everyone in the community. Places to meet new members or to ask for help about medical issues are available. Sections exist just for parents or if members want to discuss a specific intersex condition, like AIS, they can go directly there.

So that members will know that they aren't the only ones who participate, a tally of all current and even active members (those online at the time you're online) is kept and updated dynamically, showing, for example, that at the time the screen shot in figure 6 was captured there were 41 users currently on the site, more than 1,100 total members signed up. And one of those, "Glenn," was celebrating a birthday.

Being a member, which requires registration, is an important way in which a feeling of bonding is produced. It may be that not all members are on at the same time, and there is no guarantee that members will provide accurate biographical information, such as a birth date. They don't even have to provide a picture of themselves, using instead some sort of avatar or symbol if they want. But to be a member of the community, they

**FIGURE 6.** BLO discussion forum.

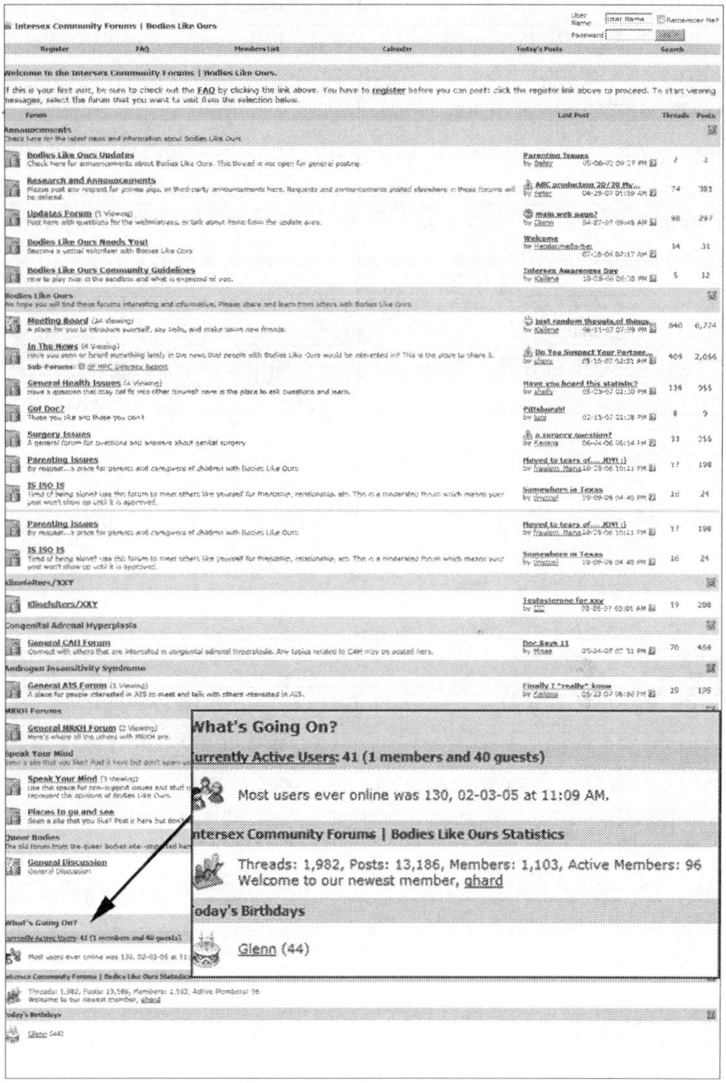

have to take the steps of registering for it. This, on some sites, means that members, once registered, get access to parts of the site that are not available to everyone. For example, the OII site allows members alone to participate in online discussions facilitated via Google Groups. Drawing a line between member and nonmembers establishes an electronic boundary of sorts that helps give shape to the virtual neighborhood. To be more than just an observer, to be an actual resident of such a neighborhood, one must be a member. Membership opens up access to information and support that nonmembers, those who have elected not to be full participants in the neighborhood, cannot enjoy. The Turner Syndrome Web site clearly indicates, as an incentive to attract new members, those benefits to members. They are listed, as figure 7 shows, on the Turner Syndrome "Members Only" page.

Is there a downside to member-only information? One obvious drawback to such member-only areas is the fact that they prohibit access. The intersex Web site discussion group or chat room allows intersex persons to communicate online in a secure, members-only environment that protects them from both public ridicule and intrusion by curious eavesdroppers. Although conducive to private, sensitive conversations, it can be argued that this diminishes the effectiveness of the rhetoric produced by intersex virtual communities. Much of the storytelling, the firsthand accounts of traumatic surgeries and, because of this, the more compelling and, thus, effective information that usurps medical truths about intersexuality can sometimes only be accessed by members. Perhaps this is why BLO's discussion forum, although it does not allow anyone but members to post, makes available even to the public its communications. Admittedly, this information may be sensitive, but making the

**Figure 7.** Turner Syndrome Society membership benefits.

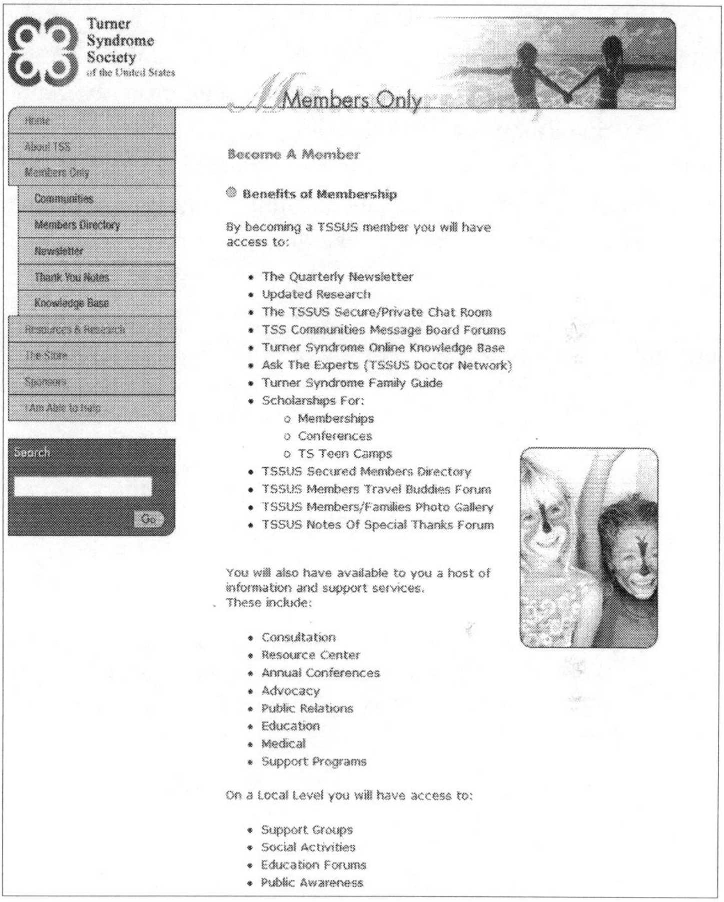


stories of frustration, pain, and anger—stories that offer different perspectives and advice—open for everyone to see means that the alternative truths they espouse will have greater resonance beyond the members of the community.

## SHARING EXPERIENCES

The BLO discussion threads found there vary, touching upon a wide variety of topics. Sometimes the conversations do touch on overt efforts to change the prevailing culture surrounding intersexuality. But more times than not, the discussions are about sharing experiences. Often, members just need help, and so they go online to places like the BLO forum because they believe there is a community of people like them, who care about them, who will listen or offer advice and support without judgment. Nancy's plea for help on June 4, 2007, titled "not sure what to do," illustrates this:

> im falling apart. i dont know what to do. i have been dealing with this for far too long. i think i lost my mind.
>
> i dont even know why i am posting here. call it a feeling. i always act on them. maybe this feeling will lead me to someone who understands.
>
> i have walked on the wire of gender for so long. falling on one side and climbing back up. falling over on the other side. sometimes i can catch my balance and just stay up there but i always fall. sometimes i can lean in the direction i feel i should fall to.
>
> it is too a point where my body doesnt exist. i live a happy normal life in my head. i force myself to have normal childhood memories. i create this whole illusion and i can live in it. but then reality interrupts and i fall so fast and the ground is far from soft. im so scared that things have gone too far to control in my head. im not always here. im not always me. im not even sure who i am right now.
>
> i dont know anymore. i tried getting help. therapy was horrible. i never want to go through that again. pouring your heart out to someone who clearly doesnt even care is not what i want. i just want somebody who understands.

> im sick of being so incredibly lonely and isolated and feeling like some kind of freak. i didnt ask for this. i dont want it. i just want to be cured.
>
> i feel this was a waste now.

Desperation like Nancy expresses carries over to other sensitive subjects, such as sex. The societal expectations regarding sex place extraordinary pressure on everyone. This pressure is perhaps more acute for those whose bodies are perceived not to meet what society expects is necessary to perform sex correctly or effectively. Men with Androgen Insensitivity Syndrome (AIS), like Stuart, a BLO poster, can suffer from genitals that are small and that do not appear to be capable of performing sex in a way Stuart believes society expects a man to be able to perform it. His frustrations with such expectations are manifested in this posting to the forum:

> To save those who haven't read the lit for what the AIS grade 3/4 manifestations are for a guy, outwardly, it means—essentially—miniature genitals. The delivering doctors (and my mom—no known dad) had to decide whether to leave me open, remove the elevated testicles, and declare me a girl with an unusally large clitoris; or sew me closed, and declare me a boy with an unusually small penis (undescended testicles to be dealt with later if necessary).
>
> What did I end up with? Testicles that are functional but non-viable, the size of large jellybeans (or maybe calamata olives?), and which rarely descend; a large, zipper-like scar from the base of my penis to my anus; and an uncircumsized, dime-diameter penis that's sewing-thimble-sized when flaccid, and about 1.5 inches when erect.
>
> I guess they made the right decision regard [*sic*] choice of sex because I turned out to be unusually athletic (for an

AIS case), look OK, and am definitely attracted to girls, not guys.

But therein also lies the frustration, because girls most definitely aren't attracted to **me.** Well, not true…initially they are, just not after they discover how I'm put together. Their reaction isn't just negative, it's, "I'm sorry, I have to go now!" And they do, hastily throwing on their clothes, never to be seen or heard from again.

So. Here's what has evolved, and why I'm posting. It's been over five years since I've even attempted to ask a woman out, never mind have sex, and a pretty good self-hatred has cropped up in the meantime. I look at those films where the guy is three times longer than the girl's hand is wide, and she can't even completely encircle it, and feel a combination of being the butt of a cosmic joke played by the Sex Gods, and pure rage at my body. So I end up hitting myself there. Legs apart, delivering blows to my pathetic "equipment" until I can't take the pain, and finally roll on the floor in tears, saying something like, "There! How do you like THAT, huh? You want to be useless? I'll show you useless!"

Sometimes I throw up, sometimes not.

Then I read—and it's what brought me here—that no matter who you are, that somewhere, somebody exists who likes/wants/needs exactly what you have. Like if a woman is as flat-chested as a boy, not to worry, there are men who find it exciting. She just has to find them.

Do I really think that finding such a girl is likely? No, but what other choice do I have than to look for one? Do I really hate my body that much? Yes. I think I would much prefer to never have been born, if I had the choice today between doing it again the way I am, and not existing at all.

Posters responding to Stuart share his concerns. Robert tries to reassure him, and his comments not only attempt show that he understands Stuart's predicament because he lives it too, but they also purposefully indict societal expectations, which although may be there but are not necessarily grounded in truth:

> I have a micro penis and no testicles too. For the first 44 years of my life I thought I was a partially developed male. I've had sex with 5 women from the time I was a teeanager [*sic*], and was married for 12 years. I may have some encouraging words. Only 1/4 of females orgasm while being penetrated - that comes directly from my certified sex therapist. The chances of you finding a woman who doesn't care how big it is is 75%.
>
> Don't give up the search. It's terribly demoralizing to be rejected and too easy to imagine it's never going to stop. But, you will find interested women. They are definitely out there.

Interestingly, despite Robert's words of encouragement, Brad, posting not soon after him, isn't convinced. As he writes,

> I am 44 and have never had a girlfriend or relationship with a woman over embarrassment and "certainty" that it would be impossible to find a woman who doesn't require more than a micropenis. So-called, open women, who claim that they are not obsessed with size say that they might even date someone with a penis as small as, say 5 inches - more than twice the length of what I have! Ugh!

Most people never give a second thought to the idea that sex is a cultural construct. They don't consider the impact of religion, of societal expectations, of a myriad of discursive forces that influence how we think about sex, who we are supposed to have sex with, and how we are supposed to do it. But intersexed persons, already marked by society as abnormal, are haunted in a way by expectation, their bodies serving as constant reminders during intimate moments like sex that they do not measure up; their penises, their vaginas, and their genitalia do not allow for "normal" sexual performance.

Such is the irony of the treatment of intersexuality in our society. We all have different faces, different noses, lips—we all have irregularities in the same basic things that we all see every day and such variance is normal. But what is rarely seen, if ever, in our society, perhaps other than by medical practitioners, our genitals, cannot be irregular or varied. Ambiguity of any kind, variance to the extreme of the imposed boundaries of normal, of that tucked away behind pants or dresses, is not acceptable.

This emphasis upon the body makes intersexed persons like Stuart and Robert feel inadequate sexually. The question is, can sharing their stories with others who empathize, who share often the same experiences, offer to them some support? Can an online community help people in their bedrooms? Such a possibility seems to be a stretch, but so much of what we know about sex is not really learned in the bedroom, at least first, is it? The rite of passage that is sex often involves stories from others, mass media vehicles such as movies and magazines, and then, at some point, the actual sex act. Yet so many speak of that act, especially the first time, being disappointing, if only because it does not measure up to what they thought it should be like. Clearly, sex is as much about our culture as it is our bodies. If there is a place where that culture can be influenced by others living with bodies and possessing sexual experiences like those who feel their own are not adequate, then there must exist the potential for such a community, even if it is online, to be influential, either by offering a home for others with similar issues regarding sex, or by presenting to communities and thus the culture beyond them the idea that intersex bodies are viable.

An example of the community reinforcing a positive image of the intersex body even in the act of offering support and bonding can be seen in the response to Pam, who posted in 2007 to the BLO forum seeking advice on surgical correction of her genitals as well as feedback on sexuality. Pam wrote, "I'm really confused about my sexuality, though. I don't feel male at all, yet I am attracted to women more than I am attracted to men. I had twice as many former girlfriends than I have former boyfriends…I wonder why? I mean—It's hard enough being intersexed without being a lesbian…Especially in my country." Dedra was one of a few to offer medical advice, indicating she should get a "cat scan." As she wrote, "I just got a cat scan to see if a uterus is hiding in me." But Jamie's remarks were not helpful in medical terms. Rather, she sought to praise the uniqueness of Pam's body:

> Sorry if it sounds a bit flippant but it sounds like you must have a really cool body. It seems like you can assume a male or a female look depending on how you're currently feeling.
>
> I'm kind of jealous of that cos I've always worried that people think I look different to the way I pecieve myself…that I don't look very female when I'm wearing girly clothes but I don't look like a man when I'm wearing bloke clothes.
>
> You can choose **either** rather than looking like neither. That could be the best of both worlds, at least superficially. I guess it's not as simple as that emotionally though. It must be really hard for those who have had surgery forced upon them.
>
> It's funny you saying about being an intersexed lesbian because when I first found out I was CAIS I couldn't help feeling I must therefore be gay based on my genes. I guess it just goes to show that nothing is straight forward.

> You have to be true to yourself in the end but as long as you're not harming someone else and you're staying safe, why not stay as you are? Sounds like you're managing to have fun?
>
> I'm sorry your parents don't seem to see it that way at the moment, I guess they're just worried about all the possible outcomes for your life. I'm sure they care though…my Mum has said she feels to blame for my intersex condition, that she gave me the wrong genes. What can I say to that? I take that to mean that she feels ashamed of me…and guilty…like I'm broken and it was her that did it. Rather than I'm fine the way I am and she's pleased with me!! see what I mean?

Jamie's words reach out, using language of her own experiences similar to Pam's so that she'll feel welcome. Then she goes further, actually complimenting Pam's body, attempting to make her feel as though it were special in a way that elsewhere is considered freakish but at BLO is valued.

## REACHING OUT—A PATTERN FOR TELLING STORIES

We see this same effort to connect with others on the subject of sexuality when Sandra writes about "Sexuality & Relationships" on the Androgen Insensitivity Syndrome Support Group of Australia's Web site. In fact, the pattern of development for reaching out to the audience is remarkably similar to the one Jamie uses, if not many other instances on intersex Web sites when members communicate with each other. First, the speakers attempt to relate their own experience to the audience, recognizing this is an entry point to the discussion as it provides a way to speak credibly on the subject, just as it offers argumentative support. Sandra writes,

> For me with CAIS, having a short vagina and no female reproductive organs has been difficult, more psychologically than physically…I feel at times afraid that I am not good enough for a relationship, that I will just do whatever I think will please the person I am with, and feel panicky at the first sign of them being ambivalent or not so interested. It's as though I feel I have to be twice as beautiful or twice as attentive and caring to make up for the feeling of inadequacy in my own femaleness.

Next, the communicators make their key points, which often involve offering advice that is again based upon lived experience, as Sandra does when she suggests that for those with AIS

> it is up to each of us to recognise these feelings in ourselves and find our own way towards healing them in whatever way we feel is right for us. I've had much success with this by doing something as simple as buying some sexy lingerie! It helped me to express my feminine side and to feel more comfortable with looking attractive. It was a big self-esteem boost!

Finally, the tendency among many intersex community messages is to offer a final line of encouragement. The communicators know that their audience is facing difficult situations and they have rarely if ever felt normal. Sandra writes,

> So to those of you who, like me, are still single and really don't want to be, don't give up hope! There are plenty of others feeling similar to yourselves, lonely and fed up with dealing with everything on their own. An intimate relationship with someone might be scary when you have to tell them things about yourself that may be confronting for them to hear, but from my own experience people are usually much more open-minded and supportive than I expect them to be. Telling your partner about your AIS can be an opportunity for both of you to

> come closer together and be a source of strength in your relationship. If you make a commitment to support and protect yourself in the best way you can, you can't fail to win!

We see this pattern again, one that reaffirms community ties, when new participants join intersex online groups. Sheila, making her first post to the Intersex/Androgynous Yahoo Group, just doesn't say, "Hi, I'm Sheila." Nor do the members there just say, "Welcome, Sheila." The exchange is imbued with lived experiences. Ironically, what have been lived experiences that have to varying degrees stigmatized them now become the means by which they establish themselves credibly in the group, and through which over time they will draw upon to welcome future new members, offer support, and reinforce community-building ties. Here is Sheila's first message:

> So, hmm my name is Sheila and I'm 24 years old, going on 25 - although I feel younger, and also look younger.
>
> I have no formal intersex diagnosis, and as far as some people are concerned, I follow a male-to-female transsexual medical regimen. I have a couple of medical opinions suspecting I have 'something' at least (as far as intersex), but they either don't have the ability, money, time, or will or expertise to have a formal diagnosis. The most likely being AIS grade 1, but I can't be certain…I came to know about intersex in April 2006, when my family doctor requested me to pass a karyotype test, because he suspected I might be intersex. I had to look it up myself, and I found that most likely was AIS, or XXY, or both—but that conditions that could look like it did not seem to fit…I started presenting as a girl full-time (only girl's clothing) on April 13th 2006 and have since. I wouldn't go back, I feel much more at ease that way.

> Finding a sympathetic knowledgeable doctor to get a diagnosis of anything has been a problem, considering I don't have income to pay specialists in the private. I rely mostly on the public sector.
>
> I'm not saying much here, trying to make a 'light post' without going too deep (because *then* I get lengthy), if you have questions, comments, anything at all, feel free to ask or say it.

Here is Kelly's response—a more established member of the group reaching out, empathizing by relating her own experience:

> Welcome to the group. Your description of doctors' reluctance to investigate any intersex possibilities is quite familiar. They seem to discourage us from considering that possibility. I learned of my extra chromosome by accident in a college human physiology class. But even then, with laboratory evidence, the doctor who prescribed my estrogen later would only do so after listing "gender dysphoria," a mental condition, as my diagnosis. After that, in an interview for a documentary on intergender people, he actually admitted to the interviewer that, many times, doctors know a child is not entirely male, but they will tell the parents the child is.

Of course, some still might perceive that exchanges like this between Kelly and Sheila, really any of the countless exchanges carried out on intersex Web sites, are not as constructive as the bonding that can and does take place in face to face (f2f) communities. "These worriers," Wellman and Gulia make clear, "are confusing the pastoralist myth of community for the reality. Community ties are already geographically dispersed, sparsely knit, connected heavily by telecommunications (phone and fax), and specialized in content" (17). Few of us even in f2f communities bond in ways that cannot be duplicated online. It also is the case that we do not separate our realities from our actions online, even if we play virtual games and take on virtual roles. We are still playing the games, and intersex persons are still bringing with them what they have experienced offline. In fact, the virtual bonds they build with each other are because of their offline experiences, making in a sense their ties perhaps stronger than those who join online communities for other purposes.

## GIVING BACK TO THE COMMUNITY

What speaks to this feeling of kinship more than anything else is the desire many members share to contribute back to the community. It is a home they wish to maintain, as the following post from Mary demonstrates:

> I can't do much typing, as my arms go numb quickly while typing, but if there is anything special you need done…I may be able to do it depending on if it is something my body will allow. I can make noise for our community, I can stand in a crowd and throw eggs, and I can write letters to the editor…Just tell me and I will try my best to perform [*sic*] my assigned tasks quickly. P.s. This site is wonderful, so, Thank-you so much for all the work you do to make it possible.

Intersex virtual community members feel they belong, and by belonging, they can be themselves while contributing to something larger. Cyndy writes,

> I guess we can be successful too (never too late) if we have the will…there really is no stopping us…and having support groups like this and wonderful people to encourage each other…who knows…the sky's the limit. Have

> I ever said that I am so grateful to God for leading me to you folks and how much I appreciate you all…well I am saying it now…never hurts to repeat it too…Love ya all.

So often feeling like they are some place where they are not like anyone else, but online in the community they want to foster, they are not alone. They are beautiful. They are connected. They know that others like them are listening and will respond. For example, when Barbara goes online the BLO discussion forum to tell others about her adopted daughter's "normalizing genital surgery" when she was in "the custody of the state," she can be confident knowing there is a community out there that will answer the question, "Any thoughts, anyone?" In fact, Kelly comes on quickly telling her "to keep doing what you're doing…Best wishes, and hugs, its great knowing your [*sic*] accepting, every child should be so lucky, we all deserve to be treated lovingly." Then Lonnie comes on, expressing the kind of outrage that one might expect to hear from a friend sympathetic to your position, indignant on your behalf over treatment that you or a loved one might have received:

> THE STATE DID WHAT!?!?!
>
> That was not the state's place! Foster care had no right to mess with her gender, not unless it was for pure health reasons, they had no right to do that! The foster care system thinks they can do whatever they want without being questioned, but that's going too far! Way too far!

Finally, Mary hears from Devanna, who offers support through, again, sharing personal experience:

> Sounds like my story. I was born in 1949 to an unwed mother at a charity hospital. I don't know for sure what

> all was done but was put up for adoption as a boy, an identity I could never come to terms with. My adopted parents were NOT open to my expression of who I was and that led to 20-odd years of WAR. It was not until I came of legal age that I could put things right and move on with my life and that created a rift with my parents that has not healed in 34 years...we know you will do right by your daughter!

It is "the accumulation of small, individual acts of assistance" like these and all the others noted previously, which "can sustain," Wellman and Gulia write, "a large community because each act is seen by the entire group and helps to perpetuate an image of generalized reciprocity and mutual aid. People know that they may not receive help from the person they helped last week, but from another network member" (9). By coming together online, by representing bodies either scarred by surgery or, if not yet conformed, that represent an in-between status not considered viable, intersex persons engage in what Butler might call a dramatic, albeit contingent, performance.

## THE IMPACT OF BONDING

Just how is this performance impactful beyond the community? Why does bonding matter? Aside from what Long and Koyoma suggest at the beginning of this chapter, that bonding enables intersex persons to reach out and feel less alone, we also should see bonding as productive beyond the individual. Hall tells us that "culture is concerned with the production and the exchange of meanings—the 'giving and taking of meaning'—between the members of a society or group. To say that two people belong to the same culture is to say that they interpret the world in roughly the same ways" (*Representation* 2). This means, therefore, that intersex persons having never visited an intersex virtual community may not know, because their understanding of intersexuality is shaped by prevailing culture, that intersex persons are neither rare nor abnormal. But once they go online and in their participation develop ties with an intersex community, their understanding changes. The culture of the intersex community they now share articulates different ideas for what it means to be intersexed. Through language, through the exchange of shared experiences, intersex persons who participate in various online intersex communities see intersexuality as something different, as not rare, as not abnormal.

The productivity of the intersex community does not stop with just its own members. Obviously, anyone, including doctors, mass media representatives, scholars, and the general public can, because of the Internet, access and be influenced by virtual intersex communities. Just as neighborhoods, even virtual ones, make themselves into what they are, Appadurai suggests, in response to wider contexts or influences upon them (186), they also, in turn, push back on those influences, shaping them. So an individual act may just be about survival, about confessing because a story needs to be told. But, as Foucault argues, "[T]here is no power that is exercised without a series of aims and objectives" (*History of Sexuality, I* 95). When intersex persons ask for help or offer advice to each other, the bonds of community that are formed also generate a culture that offers up for consumption alternative representations of what it means to be intersexed, and the presence of such meanings play a role in regulating the production of culture beyond. In fact, as we continue forward to the next chapter, we will see that the ways intersex virtual communities disseminate their culture challenge, if not rearticulate, cultural conceptions of intersexuality.

# Chapter 5

# Disseminating

The Internet allows intersex activists to make their views available twenty-four hours a day, every day. In the hands of those who understand its technology, it is extremely flexible, capable of storing and disseminating large amounts of information quickly whereas other media, other means of dissemination may be too costly (compare printing costs to posting a blog) or inaccessible. In this chapter, I argue that, in fact, the Internet does work successfully to enable intersex activists to get their messages, their views, and their stories out when traditional resources are limited.

In analyzing why people spend time on the Web, Ridings and Gefen point out that the most frequently cited reason we go and stay online is to access information. "Knowledge and information are, in general," they write, "a valuable currency or social resource in virtual communities" ("Virtual Community"). Latest trends that have seen the Internet become a more relied-upon, more credible tool for an increasingly diverse collection of users have empowered virtual communities and the information they produce. This is especially true for intersex Web sites that often deal with health information, which for so long was regarded as truthful if only disseminated by medical professionals, such as doctors, from the confines of the clinic or the peer-reviewed medical journal. Although these environments still are regarded as the ideal locales for receiving information about a medical condition, other locations, such as the Internet, are being accessed by more and more people for health information. In fact, a recent study showed that more than two-thirds of Internet users access health information online (Weiler), and of those who do go online for such information, 70% of them, according to another study in the *Journal of the American Medical Association*, use it to make important health decisions (Berland). Thus, the truth no longer exists just in the clinic. People go to Google, they visit support sites, they display a relentless willingness, a need, to go wherever they can in this postmodern world to seek out multiple providers, multiple sources of truth.

The Internet is also a sphere of communication relatively free from censorship so it enables and empowers marginalized communities to speak out, anonymously, if need be. Merely by speaking out without hindrance is discursively effective. Koyoma argues that "physicians are…afraid of our [intersex persons] presence on the [W]eb" ("Question"), primarily because online intersex persons are not isolated, not concealed, not silent, and because they can speak on issues previously addressed only by doctors whose authority went unchallenged. This power of the Internet, in fact, led two physicians, Cynthia Corpron and Joseph Lelli Jr., writing in the 2001 *Journal of Pediatric Surgery*, to argue that many intersex Web sites contain information that does not conform to endorsed protocols, and, thus, that physicians should screen sites and make recommendations to parents about which ones were suitable for viewing (1188).

## The Power of "Weak Ties"

Of course, this is to some degree a legitimate concern. On the Internet, we can now read as much as we want about any sort of medical condition. Some information is password-protected, some is inaccurate, and still some requires a certain level of knowledge to read, let alone comprehend. But as Wellman and Gulia note, people have been getting advice forever from their neighbors, often from them first before perhaps experts like doctors. "Before life on the Net," Wellman and Gulia argue, "people did not always consult expert mechanics for their cars, doctors for their bodies, or therapists for their psyches…To some extent, the Net has just made the process more accessible and more visible to others, including experts whose claims to monopolies on advice are threatened" (7). Consequently, we hear extraordinary stories, accentuating those countless other mundane exchanges that occur constantly, of people reaching out to help people, even strangers online. Rheingold writes of his online community rallying around a father who had just learned of his son's leukemia diagnosis, offering advice and support (21). Recently, in the news was the story of two women, one living in England and the other in the United States, who had never met except to exchange messages on a pregnancy support Web site. After meeting online, they shared photos of their newborns. One of the mothers noticed a strange white haze over the eye of the other mother's baby girl. She did some searching online for possible causes and came across a possible diagnosis of a rare type of childhood cancer called retinoblastoma. Although they had only what might be described as a weak, virtual tie, the mother felt comfortable enough to encourage the other mother to seek treatment from a doctor. After doing so, the diagnosis of retinoblastoma was confirmed, and the condition was treated. In the long run, this will probably mean the baby will have a normal lifespan (Considine).

Doctors, clearly, have their profession, their training, and the surrounding discourses to draw upon. Perhaps the mothers shared a common experience that was enough for one mother to trust another. For intersex activists, credibility comes in large part from their lived condition, the fact that they truly are "wounded storytellers." Still, other intersex Web sites take steps beyond this to enhance their ethos. ISNA, for example, goes out of its way on its Web site to indicate that it is governed by a board of directors, many of whom hold medical licensure or other professional credentials. This is an obvious effort to legitimize the information it presents so that it might be regarded as more authoritative, contesting in the process the "anatomo-politics of the body," part of a technique of regulation that Foucault refers to in the *History of Sexuality, Volume I* as "bio-power" (139). Anatomo-politics regulates the "body as a machine," applying norms to ensure the individual body is productive for society (139). To regulate the body, individuals are disciplined so that they are normal; they are trained, indoctrinated, and otherwise formed (physically and psychologically) so that they will be both useful and docile members of society (139). It is bio-power that conforms them so that they listen to authority and they raise compliant children who learn from their own compliance; in short, they are medically managed, their views as well as their bodies the result of "productive constraints," as Butler writes, "of certain highly gendered regulatory schemas" (*Bodies that*

*Matter* xi). But the Internet allows for the intersex person to perform an identity that challenges the regulations of bio-power. As a result, control of the intersex body through traditional knowledge of the body is challenged, and a resistant counter-discourse emerges.

The Internet can thus be seen as a vehicle that empowers intersex persons to participate more actively in the construction or, as Butler puts it, the "stylization" of their identities (*Gender Trouble* 140). As disseminators of information, they are not hidden rarities or monsters of nature but rather are authorities, producing alternative representations of themselves as well as alternative information on methods of treating intersex persons.

## Push and Pull— Libraries, Media, and Speakers' Bureau

Information is disseminated in multiple ways. Some Web sites have e-mail newsletters that interested persons can sign up for and receive. Organisation Intersex International (OII) also distributes an Intersex News Service, although to its disadvantage it is not XML (extensible markup language) driven; consequently, at this point at least, it cannot be syndicated as an RSS feed, which is something many organizations, like ISNA and Intersex Initiative, are maximizing to distribute quickly and automatically their latest news. Pushing Web content, as opposed to hoping that users on their own will be pulled in, guarantees that information will be better disseminated, simply because users don't have to go to a Web site to see its content. Rather, the content is syndicated and so can appear as news feeds on other Web sites, can even be subscribed to so that users get updates of it when information changes, both via their computers and handheld devices like iPhones.

Other methods are used beyond electronic newsletters to disseminate information. The xyTurners Web site, for example, has a section containing references to more than one hundred medical documents, divided up into categories on intersex conditions, genital surgery, XY/XO Phenotype Variability, XY/XO Prenatal Diagnosis, and medical conditions sometimes associated with XY/XO. Links are available for users to access online either complete versions of these documents or abstracts of them. The AISSG Web site also has a section with references to medical articles related to AIS. In addition, it has an entire section dedicated to providing information about relevant AIS research studies currently under way. In fact, this section not only disseminates research, but also offers the names of researchers or clinicians currently engaged in this research that it supports and wants the members of its community to support. To help solicit that support, it includes a brief description of the researcher's work and even allows the researcher to pass along a message, set off in italics, to potential supporters. An example of this packaging of information can be found in figure 8. In it the work of a researcher, Julie Alderson, is described and updates of her work are offered, as well as references to follow-up work:

OII, in addition to offering a news wire or service, provides a library that contains accessible articles written either by intersex persons or those experts who have written on intersex-related subject matter. Medical information broken into defined terms is also found on the OII site. Interestingly, visual literature is also made available so that those interested can download and use. Figure 9 depicts one example: A sticker showing an intersexed body with a confrontational caption—"Men are from Mars, Women are from Venus, I am from Planet Earth! Deal with it!"

**FIGURE 8.** Julie Alderson's research studies—AIS Web site.

**Psycho-Social Studies**

Julie Alderson (clinical psychologist), who works in the multidisciplinary team at Leeds with gynaecologist Adam Balen and geneticist Carol Chu (all of whom were guests at our Sept. '98 UK meeting), is doing a study on psychological aspects of the care of AIS patients. She says:

> *I hope to provide psychologists, gynaecologists and other medics with an insightful and authoritative account of .... what it is like to have to deal with AIS and the difficulties encountered. I am particularly interested in trying to distil, from a range of women's experiences, what it is that helps people deal with AIS (and what might hinder the process).*

Further information is available here. She is interested in talking to group members (parents and AIS girls/women) about their experience and can travel to meet you. Please contact Julie direct at Dept. of Clinical and Health Psychology, St. James's University Hospital, Beckett Street, Leeds, LS9 7TF. Tel: 0113 206 5897 (direct). Email: julie.alderson@leedsth.nhs.uk.

This study is now completed and Julie reported back to the support group at the Sept 2000 UK group meeting. The study forms the basis of the following thesis and paper:

**Alderson J: XY Women with Androgen Insensitivity Syndrome (AIS): A Qualitative Study.** *D. Clin. Psychol. thesis*, **University of Leeds School of Medicine, Academic Unit of Psychiatry and Behavioural Sciences, Sept 2000.**

**J. Alderson, A. Madill and A. Balen.** Fear of Devaluation: Understanding the Experience of Women with Androgen Insensitivity Syndrome. *Br. J. Health Psychol* (2004) 9: 81-100.

**FIGURE 9.** OII downloadable sticker.

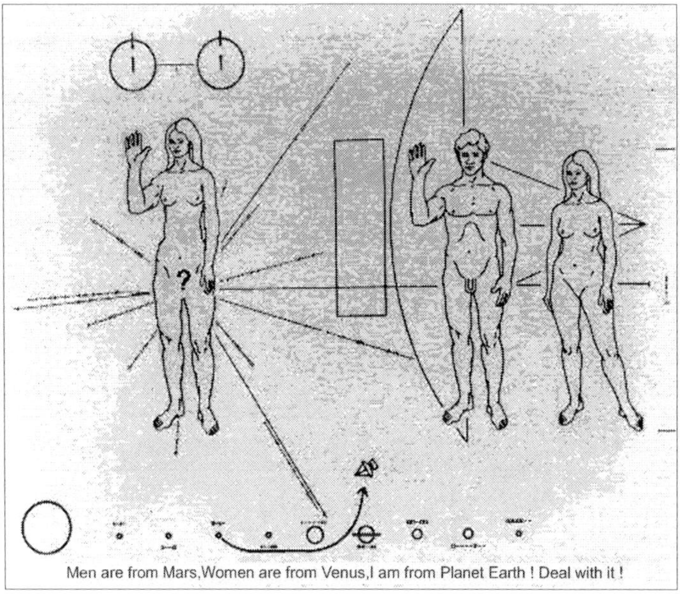

The Intersex Initiative Web site packages and disseminates information in a variety of ways as well. As its media center page shows in figure 10, it includes a section just for the media, recognizing that good dissemination, especially that which influences the mainstream, often starts and ends with the media. To make it easier for the media to report on intersex issues from the perspective of the intersex person, guidelines for writing about intersexuality are also provided and a media guide is in the works:

The Intersex Initiative site also includes links to articles on intersex issues, available in PDF format, that are written by people the site calls "intersex persons and allies." Some of the articles available are first-person accounts, while others are more

**FIGURE 10.** Intersex Initiative media center.

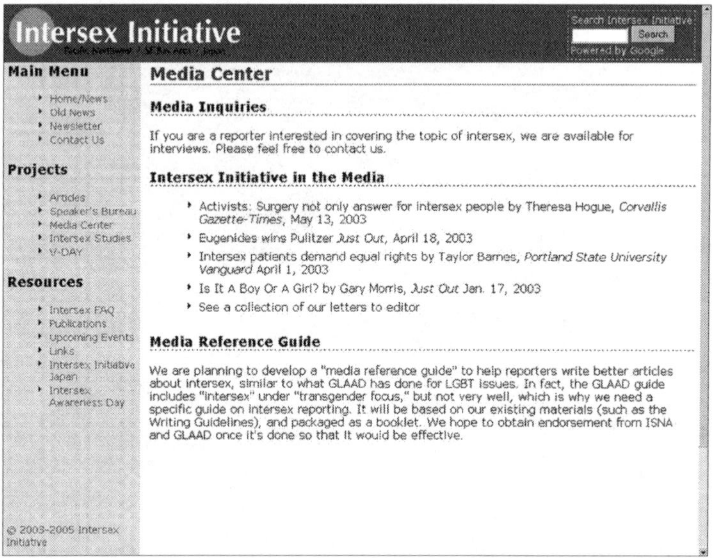

scholarly in nature. In addition to an articles section, Intersex Initiative also offers publications produced in-house, that can be downloaded as PDFs, such as an *Introduction to Intersex Activism, Teaching Intersex Issues*, and *A Speaker's Handbook for Intersex Activists and Allies*. Lastly, the site markets intersex awareness campaigns, such as Vday. Vday, inspired by Eve Ensler's play, *The Vagina Monologues*, started in 1998, and has been held every Valentine's day since. It is a successful global campaign to stop violence against women and girls. This includes rape, incest, genital mutilation, and, since 2002, thanks to intersex activist efforts, unwarranted surgeries on intersex children.

Intersex Collective, a new site just going live in 2008, advertises a "Speakers Bureau." As the site notes, the bureau represents a wide range of speakers, including the site co-founders, Long and Esther Morris Leidolf, along with David Cameron, who serves on the San Francisco's Human Rights Commission, and Jen Pagonis, an intersex scholar. What is most striking about this speakers' bureau is not that it exists. Other organizations do the same. It is that the Web is the originator of the potential for f2f communication, not the other way around. So often in the past, one of the weaknesses of Web communication and obviously, therefore, Web dissemination was the fact that it was very difficult to pull people to a Web site. Hence, the large number of expensive and arguably unsuccessful television advertisements that inventively attempted to get people to go online. But we don't have to do that anymore. Ironically, the Intersex Collective site's existence implies that we go there first. We don't call a phone number or receive a pamphlet in the mail informing us about a Web site with speakers on intersexuality. We go online, knowing that this is the place where the information we want will be disseminated. In this sense, therefore, intersex activism has benefited not only from its ability to harness the Web to build communities and send out information, but also because the Web has become a more frequented place where people no longer need to be pulled. The Web is actually pushing, sending information, even sending us, as in the case of the Intersex Collective site, back to the real world. The virtual and the real are all tied together in increasingly indistinguishable ways, and that makes communication disseminated through it not really marginal any longer—it is more mainstream, and thus regarded as more capable of generating information we will accept and use.

## INTERSEX SOCIETY OF NORTH AMERICA

ISNA stopped its operations at the time this book was being revised, but its Web presence still remains, and the information found there arguably rivals OII in its scope. ISNA's Web site distributes news through e-mail, and its main page, as the screen shot in figure 11 shows, also has news stories, offered from an ISNA perspective, that affect intersex persons:

In addition to news, the ISNA site houses a library, shown in the menu on the left-hand side of figure 11, that contains both an online bibliography and another version that is so extensive that users need to access it via special bibliography software, EndNote. Information on relevant legal cases is collected in a law section. Each case is described and summarized from an intersex perspective. Furthermore, various books and videos created by or about intersex persons are reviewed. Finally, a "What We're Reading" section is available. In this section, ISNA staffers or board members tell users about the documents they are reading that meet with their approval. In short, ISNA offers up news, research, videos, historical documents, legal case summaries,

**FIGURE 11.** ISNA Web site home page.

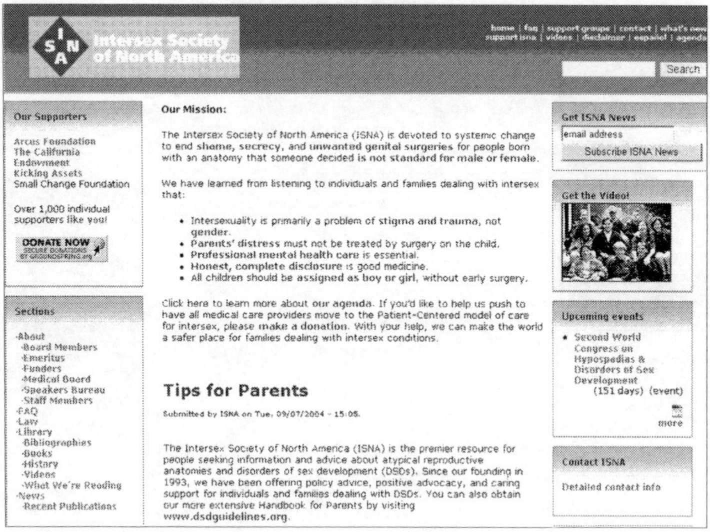


and announcements of upcoming events. These can be printed or delivered via e-mail.

## Organisation Internationale des Intersexes

OII's virtual neighborhood lays claim to being the home for the largest intersex community, which is an attempt obviously to enact credibility, since, if there are so many people already members, the information found there must be useful and reliable. Certainly, OII has purposely constructed itself so that it reaches out to the largest possible number of intersexed persons. As the screen shot shown of its home page in figure 12 illustrates, OII is an international organization, actually based in Quebec, Canada, not the United States. In keeping with this

**FIGURE 12.** Organisation Intersex International (OII) home page.

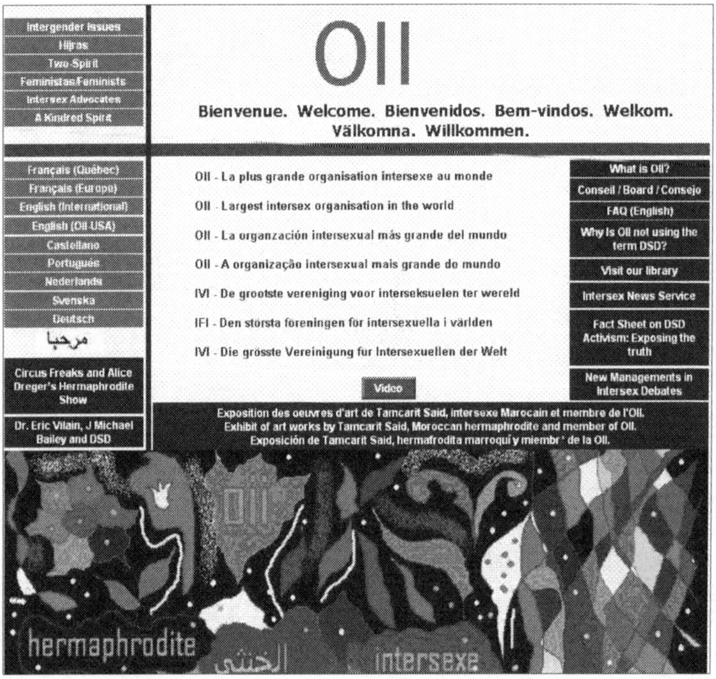


international emphasis, OII disseminates its information in multiple languages: English, French, Portugese, Castellano, Dutch, German, Swedish (Svenska), and Arabic.

Its operating board is made up of people of different nationalities, and beyond the main home there are language as well as region specific (such as OII—USA, shown in figure 13) sections with their own look (see the OII—USA Web site that follows) that, whereas performing the same mission of the larger OII virtual neighborhood, cater specifically to the needs of their participants.

**FIGURE 13.** OII's USA home page.

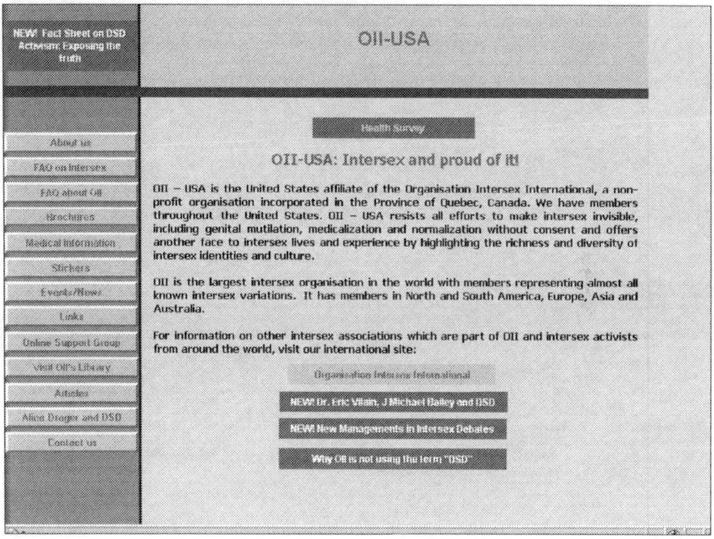


OII's online presence ideally illustrates Appadurai's argument that global flows of people, of their ideas, of their norms, have unmoored culture. Via the Web, those who might not ever find a sense of community, either because they do not know that there are others like them or that the language of the support offered to them is not their own, have reason, and the means, to come together. They can visit and participate in discussion at Intersexions Café (see figure 14) where conversation is private (members only can access) and available in French, Spanish, English, and German.

OII supports and it asserts. It makes available its official positions on health care and third sex discussions regarding intersexuality. It displays art by intersex artists. It advocates for Intersex Solidarity Day. The OII community is a place where

**FIGURE 14.** OII's Intersexions Café.

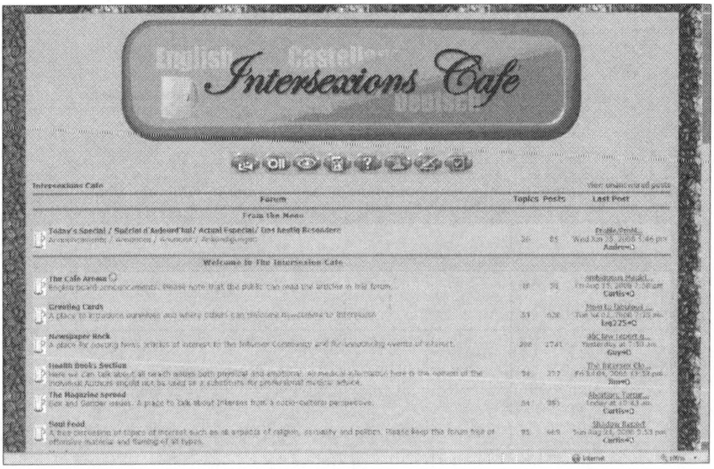


intersex persons can belong, and the impact on the rest of society because of such belonging is that a large community with shared interests has a reach that stretches past its own self-generation.

## Recommending Doctors

Still another kind of information disseminated by certain intersex Web sites, like OII, is a list of recommended doctors and clinics. AISSG, for example, not only offers listings of preferred medical practitioners throughout the world, but it also provides guidance to doctors on how best to communicate with, to treat, intersex persons. As figure 15 shows in the "A Message to Medics" section, practitioners are forewarned that they need to be open with patients to be able to be recommended, primarily because intersex persons (specifically, AIS patients) "have been extremely stigmatized" by their treatment, including how their

**FIGURE 15.** AISSG-recommended clinician information.

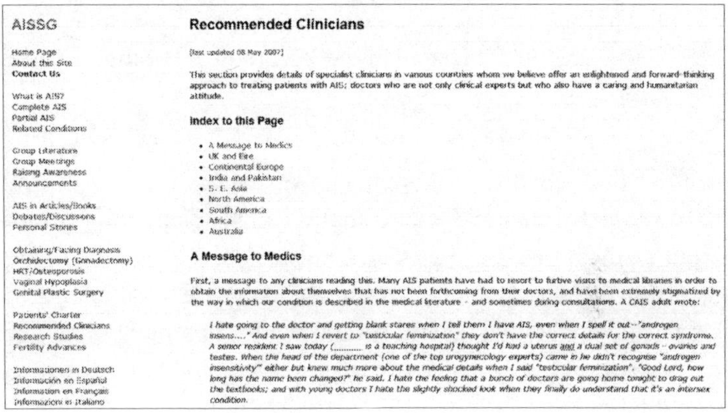


conditions have been described and how tight-lipped doctors have been in consulting openly with them about treatment:

MRKH's doctor list is available as a PDF file. Recommended doctors are listed with contact information, their state or country of practice, and brief commentaries from previous patients. One doctor on the list, for example, Michael Keating, a urologist from Florida, is described as having an "excellent bedside manner" (2). AISSG, too, lists doctors from around the world, including North and South America, Europe, the United Kingdom, Australia, and India. Like those found on the MRKH.org site, they are listed not only with their names and contact information but also with brief stories that explain why they deserve the recommendation. For example, by clicking on the North America link, users see Dr. Bruce E. Wilson, MD as the first entry. He is described and thus endorsed as follows:

> Dr. Bruce Wilson is active in trying to change the way intersex conditions are dealt with, has contributed a chapter to Alice Dreger's book *Intersex in the Age of Ethics* (see

> Books and Articles), and has been very helpful to one of his former patients (a member of our group) who contacted him aged 27 (as reported in ALIAS No 16). He attended the Aug 2000 meeting of our US group in Boston as a guest speaker.

Such recommendation lists and their details not only demonstrate the capability of intersex Web sites like AISSG and MRKH.org to match the information that adherents of traditional medical discourse produce, but they also serve to counter the power held by doctors and clinics to decide what is best for patients, and who that patient should see for treatment or advice. It is not a doctor's organization or any other type of professional accreditation, after all, that serves to recommend the doctors on the intersex sites. Rather, recommendations are based on intersex activists' determination of how doctors have treated intersex patients or dealt with the subject of intersexuality in the past.

To date (as of October, 2008), there have been more than 16,000 posts made just on the BLO discussion forum. That's thousands of stories, thousands of experiences told, of pieces of advice asked for and given. Thanna demonstrates just one when she posts, "This is so hard to ask. The Subject matter is sensitive. Is it normal to only have a vag. [vagina] that does not grow beyond the size of an infants [*sic*]? If one has surgery will you lose sensation?" Courtney responds:

> Pressure dilation is the way to go. Most often have a lot of success with this method and it is definitely the way to go before even considering having vaginoplasty performed which is bound to reduce sensitivity to some degree or even possibly make it too sensitive as I've recently learned from others.
>
> As for the clitoris being small, at least in AIS, I've never heard of that one. In PAIS grades 6–4 it's actually usually oversized, grade 3 would usually either be

> considered an oversized clitoris or perhaps micro penis with severe hypospadias. I really don't want to disagree with Miriam, it's just that I've personally never heard of it being undersized before. With complete AIS, there would be no pubic hair (or extremely little) giving the vulva a juviline apparence but it would be fully developed except of course for the vagina being blind.

This exchange, like so many of the others, does not signal just that a community-building tie has been made. Hall writes, "The subjects of the local, of the margin, can only come into representation by, as it were, recovering their own hidden histories" ("The Local and the Global" 34–5). Therefore, in uncovering information and sharing it through a variety of means, intersex persons talk to themselves and also the rest of society "from the bottom up," Hall writes, "instead of from the top down" ("The Local and the Global" 35). Their stories don't necessarily have a known history, an accepted folklore to articulate their meanings. They must do that for themselves, one story at a time. As they are disseminated via the Web, that happens, and as these new stories emerge, they resonate productively, confronting, as the next chapter shows, cultural assumptions.

# Chapter 6

# Confronting

Many intersex Web sites post stories told by intersex persons recounting the pain and shame of medical treatment, and of the first time they learned of their condition, which often did not occur until adulthood. They speak of how others treated them in school, as Anna's story shows us, and of how they still struggle with relationships.

## Dave and Christine's Story

One such story, posted in August, 2003 to the AISSG site, is from Dave and Christine. Their story offers a parents' perspective on the traditional treatment of intersex persons that directly confronts the idea that parents can better handle the reality of raising an intersex child if they are only given limited information and if they are encouraged to participate fully in keeping the condition secret from their children:

> Joanne was born in 1972; she was a beautiful little girl. Whilst in the maternity hospital I was moved to a side ward and told to put my washing for the baby in a bag marked contaminated linen. I had no idea what the reason was despite asking over and over again. I was sent home early and had an appointment with a very nasty doctor who didn't tell me anything about Jo but kept asking me if I was going to have any more children.
>
> A few weeks later we saw another doctor who was much nicer, but on reflection didn't know too much. He told us Jo had a condition called AIS, which meant she was born with testes instead of ovaries…
>
> Jo had her testes removed when she was fifteen. We were advised to tell her that she was having a hernia operation because she had been born with a hernia, which is how they detected her condition. My husband and I will never forget Joanne's face when the doctors told her the hernia had strangulated her ovaries and that she would not be able to have children. This was the advice we were given by the medical profession…
>
> Looking back over the years, the deceit was dreadful and the dealings we had with some of the medical profession was disgraceful…
>
> I still remember a time when Jo was in hospital as a baby and the nurse said to us "isn't it ironic that she's in a blue baby-gro." And when Jo was a teenager she had laryngitis and the GP said to us "perhaps her voice is breaking." These comments were made from people in the medical profession, people whose advice and judgment you're supposed to trust.
>
> We just hope that there is no more deceit for AIS children being born today and that the medical profession

has become more educated and sympathetic towards the condition.

This story about Joanne makes clear the problems intersex activists have with the treatment of intersex children, and the misinformation given to them and their parents. It is further confrontational because in witnessing to others online, Joanne's parents do not confess their experiences solely to a doctor. Power, as Foucault notes, rests with the one who listens to the confession rather than the confessor. Both the confession and the examination, then, are tools of power meant to normalize the individual. If, through the confession or examination, the examiner deems a person abnormal in action, words, or appearance, the examiner has the power both to declare that person abnormal and, on behalf of society, take action to normalize him or her.

## Subverting Authority—The Power of Confession

However, when intersex persons tell stories online, confessing the details of their treatment to everyone who might sympathize or empathize, then the power of the medical practitioner as examiner and confessor is subverted. Power is drawn from the practitioner and distributed to everyone participating in the act of confession. It is not that power is seized completely from the doctor. One culture, providing signification for what it means to be intersexual and how intersexuality should be treated, is not destroyed to be replaced by another. "Web activism," Michael Dartnell writes, "does not threaten the state or its ability for autonomous action" (10). But activism instead "transforms the contexts in which states operate" (10). In other words, Web activism enables the intersex person to become more active players in contributing to the creation of meanings within culture. We see this at work in the BLO discussion forum when a member, Fredrica, requests help, not from doctors, but from other community members in locating a doctor for her child:

> Having had a consultation with my children's endo this morning, it appears that my daughter (7) needs to see a paediatric urologist fairly urgently. Since birth she has had persistent nocturia (bedwetting). Psychological causes have been ruled out. She also had mild vaginal adhesions as a baby and toddler which seem to have sorted themselves out without any intervention…I am concerned that any potential surgery is carried out by an IS friendly doctor and not someone who might believe that a largish clitoris ought to be reduced…If anyone here at BLO has some friendly advice I would be most grateful to hear it.

Almost immediately after her posting on the same day, March 25, 2005, Fredrica receives a response from another BLO member, June. This response, seen in the following quote, is rhetorically subversive not only because it is an intersex activist who provides advice on a medical issue, but also because it is the seemingly "knife-happy" actions of the doctor, and not the body of the intersex person, which are characterized as monstrously brutal:

> Well, I can tell you to avoid at all costs Great Ormond St. The reason for this is based upon comments by doctors there calling us things like "green-wellied loonies" "Us" is intersex advocates, btw. Another doctor there (98% sure he is from there) said the birth of a newborn with intersex is as bad if not worse than having a stillborn baby…I would contact Sara Creighton and see if she can recommend someone. Just remember that every problem to a surgeon is one that requires a scalpal [*sic*] to fix.

In this simple exchange between a parent and an intersex activist, the power of the clinic, of that rhetoric which holds only doctors or other professionals can listen and provide advice, is thwarted online. It is not that the parent of the intersex child intends to forego doctors. But, instead of going to doctors only for advice about doctors, the parent goes online. The Internet, in this case and so many others like it, enables people to find alternative advice, and it enables the intersex activist to give it, to confront rhetoric with rhetoric.

## The Missing Vagina Monologues

Other methods of confronting the practitioners, research, agendas, and protocols of traditional medical discourse are also at work, sometimes reemploying what are tools of resistance that have been used in other circumstances but have usefulness beyond those contexts. An example is the MRKH.org Web site's (figure 16) use of the popular play, *The Vagina Monologues*, to further its efforts at drawing attention to issues of concern to MRKH women.

Fittingly, the original *The Vagina Monologues* is a vehicle for subversion, designed through its insistent focus on the vagina to demystify the societal taboos surrounding it. By employing the structure of *The Vagina Dialogues* to create *The Missing Vagina Dialogues*, Esther Morris, its author and the founder of MRKH, attempts also to demystify and perhaps thereby educate. In cases of MRKH, women afflicted lack, for congenital reasons, a vagina. They undergo great scrutiny from physicians, and their missing vagina causes both psychological and physical problems. Morris obviously decided that if *The Vagina Monologues* can attract notable attention to a subject regarded as taboo, perhaps *The Missing Vagina Monologues* can do the

**FIGURE 16.** MRKH home page.

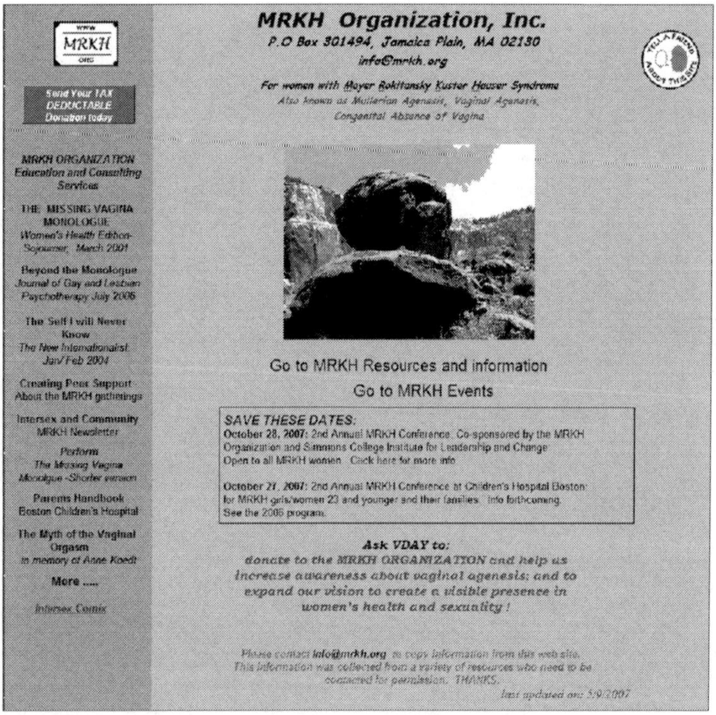


same thing for MRKH and those it represents. An excerpt of *The Missing Vagina Dialogues*, written in much the same way as the play it usurps, is as follows:

> *Curious to learn what women were talking about I read the "Vagina Monologues." I wasn't curious about the play. I was curious about vaginas. I am a woman who was born without one.*
>
> My life completely changed when I was 13 and sent home from camp with abdominal pain. When I was examined

> they found that I had no vagina, and they could detect no uterus…my doctors recommended vaginal reconstruction so I could have a normal sex life with my husband when I got married. I never had a chance to want a vagina I simply had to have one. I went from selling Girl Scout Cookies to correcting my sexual dysfunction in one afternoon. I spent the next few years going to specialists, having tests to confirm my gender, being probed by curious doctors and interns with multiple instruments in multiple holes at multiple times. They saw their definition of normal, and I wasn't it. Like an android on an assembly line, I had no concept of feeling that this body was mine. But I was not yet capable of wondering who was having the biggest problem with my body. Was it me or the people treating me? There was so much focus on the woman I "should be" that I lost all knowledge of the girl who I was. Wishing I could have had choices…The chance to desire change on my own—embracing opportunity rather than fearing who I was supposed to be.

Morris encourages others with MRKH to perform the play, posting it in its entirety on the MRKH site. Like the play it usurps, its discussion of the vagina, or, here, the lack of one is a means of interrogating the societal expectations that led to the author's own treatment and others with MRKH. It is also revolutionary, asking questions as a way of engaging that treatment. These are questions Morris thought she had to face alone because she was told by her doctors that she "would never meet another woman such as myself." Of course, she did meet other women, other intersex persons, like her, and the questions she asks as follows, posted online and framed within the structure of a play, are ones shared by other intersex persons who also resist, either by reading this play or participating in Internet intersex communities:

> Why was my gender challenged in the first place, and then confirmed like something I didn't already know? Why

> was my body taken away and rearranged like a sexual Action Figure by men with knives? What was the need to feminize my body, which actually neutered my soul? The most important discovery I made was to learn that there are many variations of gender. This wasn't just about bodies - it was about genocide of variation.

## OII Special Investigation— "Deconstructing the Feminine Essence Narrative"

Confrontation is not just a function of performance, such as the aforementioned play, or narrative as the many stories illustrate. It is often manifested in rules, as the OII site demonstrates with its "No Normalization Zone" policy. Because, as the policy notes, "[W]e in the intersex community have been so normalised [*sic*] against our consent," there are no limits placed on what is or is not acceptable material for discussion. Rather, the policy states, "[P]eople are allowed to be controversial, to disagree, to simply be themselves and state how they feel about intersex without worrying who might take offense, including other members of OII." Obviously, implicit in this rhetoric is the belief, borne out by much research, and that the 2000 AAP statement illustrates, that intersex persons have never had a voice in their treatment. So a Web site run by and for them will not ignore or censor any of their opinions, even if they are controversial enough to be questioned.

Questioning, in fact, as directed through investigative work, is at the heart of another example of confrontation found on the OII site. As the largest intersex community, OII regards itself not just as a home for discussion or a resource for information, but also a watchdog, tasked with exposing and debunking any ongoing research that it regards as negative towards the efforts of intersex persons to be normal, to pursue an agenda they control. In what is called an "ongoing OII investigation,"

Curtis Hinkle, the OII founder, introduces a multi-part look at "Deconstructing the Feminine Essence Narrative." In both tone and purpose, this report found online at the OII site is confrontational. Hinkle writes that OII is "working with an informant who has primary sources and e-mails from the individuals involved in an attempt to deconstruct the 'feminine essence narrative.'" The requirement for an informant suggests that one must work undercover, that the examination being conducted is rhetorically subversive, meant to undermine an operation that is intent on covering something up.

That operation, or position, is one that Hinkle recognizes might be considered "tangential" to intersex persons. But it involves, according to Hinkle, the "Current protocols for the management of intersex children," or the paradigm long adhered to that pushes for "GENDER assignment as soon as possible" if a child is born intersexed. The problem, Hinkle points out, using language similar to that of other intersex activists, is that "intersex children are assigned a GENDER, which is expected to reflect their gender identity, and many are assigned a "FEMALE GENDER," which is nothing more than an expectation that their lives will reflect a 'feminine essence narrative.'"

In the report, Hinkle suggests that a number of experts, many of whom are regarded as advocates for intersex persons, are engaged in a cover-up of the feminine essence narrative. Hinkle provides documentation from other experts who call into question efforts to ignore the fact that gender assignment is still occurring, meaning that intersex persons lack a voice in determining their identity. Further, Hinkle argues that trends toward making intersex into DSD, or disorders of sexual development, "will create a much larger class of people to pathologize as paraphilic (or suffering from a fetish)." Intersexuality, in other words, the investigation concludes, will become further cemented as a pathological abnormality, something requiring even more doctor-controlled investigation and treatment.

Fueled with examples, studies, insider information, and an electronic platform to speak from, this investigation ultimately is about confronting any efforts to hold on to past methods for the treatment of intersexuality, as well as any trends that do not put the intersexed person first. As Hinkle writes, "We need to have a new paradigm. Treatment needs to be PERSON centered first, and CONDITION centered, second. At present the DSD paradigm is CONDITION focused—not PERSON focused."

## Challenging, Advising, Convincing Doctors

Returning again to the AISSG Web site, we also see an indictment of medical practice and research found on the OII site. AISSG-provided information on medical procedures comes in the same format in which one would expect medical information, but from perspectives on treatment that would rarely if ever appear in a doctor-produced document. For example, the information AISSG provides for gonadectomy, or the surgical removal of testes, begins with a definition of what it is, and also includes a detailed description of how it is performed. However, rather than assuring the patient that recovery is easy or describing symptoms typical of normal recuperation after surgery, AISSG includes opinions from AIS patients who have undergone a gonadectomy. Typically, doctors, adhering to traditional medical management protocols regarding AIS, urge that AIS patients undergo gonadectomies at the earliest age possible. Assuming that people visiting the AISSG page are doing so because doctors have encouraged them or their children to get a gonadectomy, AISSG directly counters presumed medical advice by including information from those AIS persons who have undergone the surgery. This information would rarely, if ever, be made available to a patient or family by the doctor. AISSG not only argues that testicular removal is unnecessary, but that many survivors of the surgery have had difficulty adjusting after the procedure:

> We [AISSG] have one or two group members in their 30s/40s who still have their testes intact and are perfectly healthy. We also have a very significant number of members who say that they never felt the same after having their gonadectomy, had great trouble getting used to HRT, finding a suitable preparation etc. Some group members are very angry that they had their testes removed without their permission.

What also rarely, if ever, appears as part of a doctor's advice to a patient, is doubt surrounding the assumption that underlay the recommendation for surgery, let alone doubts about possible negative outcomes. Surgeons expect to speak from an authoritative position, and to have their authority respected even about subjects they are not necessarily authorities on; therefore, they tend to sound confident about the procedures they recommend. To demonstrate that doctors are not always right or sure of their recommendations, and thus to counter medical authority, the AISSG site includes a story at the end of its discussion on gonadectomies. The story relates a meeting in which a number of pediatric urologists, appearing before a group of AIS patients, were so unsure about how to treat AIS patients that they asked for advice on how to do it. As the story goes, "[T]here was a general air of anxiety amongst a number of paediatric clinicians. Several were getting quite agitated and saying things like: 'Well, I have the parents in front of me…' or 'I'm in the hot seat…' 'and they are asking me what I advise, so I have to tell them *something*...so what do I tell them (e.g. about early/late gonadectomy)?'" Framing the AIS persons, rather than the doctors, as experts, and, as a result, usurping the discourse that has empowered doctors as authorities, the AISSG discussion then reports that the AIS persons turned the tables on the doctors, using the authority of their experiences living with AIS to answer their questions:

> 'What you do is this...you tell them that the risk of gonadal cancer, in AIS at least, is very low before puberty/early adulthood...you put them in touch with a clinical psychologist if possible, and you put them in touch with other parents via the support group...so they can discuss things at length, and in their own time, and can explore all the angles. There is no hurry to do surgery unless it's a life-threatening situation.' Intersex is largely a psychosocial phenomenon and not a medical emergency, and doctors should not feel they have to be the only people to advise parents.

As this example illustrates, intersex activists use the Internet to privilege their information, and this enables them to function as experts, not because of training, but because of lived experience. This tactic has been effective as can be seen in the revised AAP statement that now recommends a more active role for support groups, made up of intersexed persons or their families and advocates, to play a more substantial role in providing counseling to the parents of recently diagnosed intersex children.

I remember researching intersexuality in the years leading up to the revised AAP statement. At that time, one could easily see the growing effectiveness of intersex activism online just by conducting a simple search of an intersex-related subject using the popular search engine Google. Intersex Web site administrators had so adeptly usurped the technologies of the Internet that when I searched for "American Academy Pediatrics Intersex," information produced by intersex Web sites about the American Academy of Pediatrics (AAP) and its stance on intersex treatment appeared even before the AAP's official position. In fact, the AAP home page appeared so far down that on some computer monitors it could not be seen without users scrolling down to it.

Obviously, there is tremendous value in being at the top of the list of search engine returns, which is why many organizations pay a premium to be placed there. Users tend to click on the first few items only. If they had done this, as I'm sure they did, for the search query in question regarding the AAP and intersex, they would have more than likely gone to intersex Web sites where the rhetoric of those organizations, not AAP, would have shaped discussion of intersexuality. In fact, at that time, the first link was directly to ISNA's site, which provided commentary on the AAP official position circa 2000. The second link was actually an article written by Alice Dreger, at that time ISNA Board President, which critically evaluates a urologist convention in San Francisco. Any reporter or member of the general public, any medical professional, any intersexed person or parent of an intersexed child, would have had the information they found colored by the rhetoric of intersex activists first, allowing them, therefore, to be the first to shape the context as well as the debate surrounding intersex treatment.

One need only visit the Johns Hopkins University Web site to see this. In an article, "Into the Hands of Babes," written for the online *Johns Hopkins Magazine*, William Reiner, a Johns Hopkins doctor and former proponent of the aggressive surgical management of intersex children, now adopts a position completely at odds with the positions taken by Johns Hopkins (which had sponsored and adhered to the recommendations of John Money) regarding the management of intersex children. According to the article author Melissa Hendricks, Reiner now believes, due to his own research and interactions with intersex activists like Cheryl Chase, that "if a child has ambiguous genitals…families and physicians might decide using the best medical information available whether to call them boys or girls, but wait for the child to decide whether to opt for genital surgery." Hendricks further writes that, like Reiner, "[O]ther physicians and psychologists who have treated [intersex] patients…say they, too, have changed their views."

Of course, these changed views cannot be solely attributed to the presence of intersex Web sites. But Dreger makes a useful point when she writes that doctors, like everyone else, "change because they hear in the news that they should" ("Question"). Thanks to the Internet, intersex activists have an affordable, available platform from which they can help shape the news that doctors hear. In fact, the Internet enables them to thwart the prohibitive power of medical authority to restrict discourse. Perhaps the Internet does not allow intersexed persons to speak truly and completely freely. As Foucault states, "[W]e are not free to say just anything, that we cannot simply speak of anything…not just anyone, finally, may speak of just anything" (*Archaeology* 216). However, it does allow them to disseminate their positions on intersex medical management and treatment. Via the Internet, intersex persons are not abnormal, not isolated, and not, importantly, voiceless. They also are not so easily punished for failing, Butler states, "to do their gender right" (*Gender Trouble* 140).

Obviously, multiple problems still exist that make the fight challenging and that keep the rhetoric still ongoing. For one, many intersex people are still isolated by their treatment, or sufficiently stigmatized that they do not wish to come forward and lend their voices to establishing new truths. Second, although the Internet is an alternative growing in popularity, it still cannot adequately compare to the clinic or hospital in terms of dispensing medical-related truths. Finally, the "will to truth" that Foucault discusses in *The Archaeology of Knowledge* relies upon powerful rules of exclusion that are not easily countered. Unlike other discourses where multiple roles have been privileged to speak, medical discourse has long privileged just one, the doctor, and many apparatuses are now built up to support that. So many, in fact, that even for many intersex persons themselves it does not make sense to seek the truth from anyone but a doctor. According to Foucault, this is because the will to truth is immersed in what we regard as true, that we cannot tell the two apart, cannot see that a doctor's authority, a doctor's gaze, a doctor's truths are all creations of discourse:

> True discourse, liberated by the nature of its form from desire and power, is incapable of recognizing the will to truth which pervades it; and the will to truth, having imposed itself upon us for so long, is such that the truth it seeks to reveal cannot fail to mask it. (219)

Still, the Internet's rhizomic nature, the freedom and accessibility it affords, and the fact that more and more people use it even for health knowledge—all of this has already enabled intersex activists to form communities online, and the increasing number of these sites, as well as the additional critical works focused on intersex issues (including this one), offer further proof that the truths of the medical discourse regarding intersexuality are being increasingly questioned. As Gilles Deleuze and Felix Guattari write, "A rhizome ceaselessly establishes connections between semiotic chains, organizations of power, and circumstances relative to the arts, sciences, and social struggles" (7). The Web, therefore, cannot be shut down like a printing press. The information on it doesn't run out of money for advertising. We see even that when organizations stop functioning, like ISNA, the ideas continue on. "A rhizome may be broken," Deleuze and Guattari argue, "shattered at a given spot, but it will start up again on one of its old lines, or on new lines" (9).

The growing number of doctors lending their authority, as the online intersex recommendation lists illustrate, further demonstrates this. Before the explosion of the Web, especially social networking beginning in the early part of the twenty-first century (2000 on), intersex activism was beginning to become effective. But the Web, as an effective, credible, affordable, online tool for activism, has played a key role in charging this activism through the community-building it enables, and the confrontational rhetoric that community is capable of disseminating. The change is not uniform. Cultural production is ongoing, and as du Gay et al. points out, "the processes of production," such as we've just examined taking place via intersex virtual communities, "only provide a series of possibilities that have to be realized in and through consumption" (59). But evidence does seem to show that the clinic, once home to a protocol for treating intersexuality that stigmatized thousands of children, is a locale where new trends in the treatment of intersexuality have been consumed. At Seattle's Children's Hospital, for example, staff has compiled a collection of grand rounds regarding what is called a "post-modern approach to intersex." Interestingly, these grand rounds are available online, streaming Web videos on a variety of related subjects from physicians and other professionals practicing there.

As is often the case with the clash of rhetorics, one has not destroyed the other. Rather, one has made its mark on the other.

Through confrontation, a new Age, far removed from the Age of Gonads or of Conversion, has begun to take hold, and a new rhetoric, such as the one espoused by the revised AAP statement, by the Seattle Children's Hospital, by a host of other practitioners, speaks new truths.

# Chapter 7

# Conclusion

This book's focus has been admittedly limited. Even in its rhetorical analysis of intersex Web sites, it has not looked at every site, but rather focused on a select few in an effort to explain how intersex virtual neighborhoods create alternative rhetoric to support ongoing intersex activism. Although called for, such an in-depth study until now has not been completed. Other studies certainly could build from my contribution here, and I would encourage it.

## DSD Controversy

The fact is that intersex activism and the rhetoric that surrounds it, online or off, remains unsettled and requires additional discussion. Recent controversy surrounding the use of either intersexuality or DSD, one that Curtis Hinkle references in the previous chapter, illustrates this. A very public debate, much of it occurring online, has played out as intersex activists have argued back and forth over how best to designate their status. A series of electronic letters submitted in response to the 2006 AAP Consensus Statement, where DSD, as a replacement for intersexuality, was first forwarded, highlights the debate. Some, especially clinicians like Durval Damiani, offer praise and support for the new terminology. Others, such as intersex activist Lynnell Stephani Long, express support for DSD as a tool for communicating with parents and with doctors regarding children just born with an intersex condition. She writes, "[I]f we talk in terms doctors can understand, they seem to understand us better." But she also sees intersex as "a badge of honor." In other words, DSD has value for her has a role in communicating with doctors, but intersexed is who she is and by using it she is empowering it less as a clinical term and more as a political statement about the history of her treatment, about the culture responsible for it, and about her own efforts to embrace her experiences and fight against them and for others like her.

The phrasing "disorder" bothers others, even if it is meant to be used to speak with doctors in language they understand. This, in fact, has led Milton Diamond to argue for the terminology, Variations of Sexual Development, or VSD, not DSD. According to Diamond and Hazel Beh, "[V]ariations of sex development inextricably involve sexual and gender identity, sexuality, and one's innate sense of self. Thus, the term disorder is far too narrow and too pathological to be accurate."

## Cultural Change Never Uniform

What term is the right one to use is of less interest to me than is the nature of the rhetoric involved. Intersexuality is a medical term co-opted, undermined, redefined, and now debated, and this is done because every term, regardless of who coins it or who has the authority to define it, is a product of culture, and as a product of that culture is subject to manipulation, both by those who are supposed to apply it and those who are not. The controversy surrounding its definition, even how it should be named, illustrates its social construction and the struggle that occurs, one point at a time, in the constant flux that is that construction. What it also shows says something, however, about the changes that have occurred in culture. No longer are just medical professionals deciding what intersexed persons should be called and how they should be treated. The debate demonstrates that intersexed persons, although they may not agree, have a voice in determining their designation, their authenticity, or their truth. Writing in the book, *Questions of Cultural Identity*, Stuart Hall asserts that our identities are "points of temporary attachment to the subject positions which discursive practices construct for us" (6). In other words, they are produced representations of who we are supposed to be, what we can be within the constraints of discourse. But what we can be, what the intersexed person can be, is not set in stone. Over time, identity has been rearticulated, and as the culture changes, the identity changes, just as the changing of that identity, the contributions that come from it, change the culture.

This is not a uniform process of change. According to Kuhn, the nature of paradigmatic thinking that scientists, doctors included, "attempt to force nature into the relatively preformed and inflexible box that the paradigm supplies" (24). In fact, Kuhn further asserts, "[N]o part of the aim of normal science is to call forth new sorts of phenomena; indeed those that will not fit the box are often not seen at all" (24). But, "[A]s a direct result of intersexuals' vocal critiques," Preves writes, "clinicians have started to reform their practices—and their frames of intersex and intersex treatment—and many are now claiming to be far less eager advocates of surgical intervention of intersexed infants and children" ("Out of the O.R." 209).

## Sharon's Story

Nor has the change been fast and all-encompassing. And not that it should be. As Iain Morland argues, perhaps the most ethical thing to do with regard to intersexuality "in this historical postmodern moment…[is]…no longer rush to pronounce the single right way to manage intersex, but admit uncertainty, replace dogma with discussion" (331). Sharon, telling her story on the BLO forum, adds to this discussion. Where else could she tell such a story and have it read by hundreds if not thousands of others, many of whom have similar experiences and frustrations she expresses?

> Dear Friends,
>
> I feel I must apologize for not writing here in the forums over the last number of months. I needed a break as I have recently come to the sad and horrific realization about my physical state of being due to a test that was accomplished in late September, and recent additional physical discoveries.
>
> The test was a MRI. The discoveries were additional scarring from my IGM. The MRI indicated a "normal male pelvis." The scars consisted of a 4 in one on the left thigh, and another was where my vaginal lips were sewn together. The thigh scar was due to the harvesting of a small length of artery and vein to allow the removal and lowering of my ovo-testis from it's connection to my uterus/fallopian tube, into the scrotum, after removal of

the uterus, cervix, and upper portion of the vagina. The vaginal scar was made to look like the central ridge of the scrotum, but due to recent stretching, the suture scarring has now become visible.

I was born a feminine 'true' hermaphrodite with one testicle, one ovo-testis, and a Y chromosome, that was surgically made to look like a boy. But that is all I will ever be – a boy. Not a man – as I have generally feminine tissue due to my PAIS – but a boy. I do not look like the "50 y.o. Man" that I'm 'supposed' to be. I'm physically more akin to a FtM trans person that has undergone 30 years of testosterone poisoning. Since the advent of my recent female puberty, I've lost close to 40 pounds, and my torso has become significantly feminine, but my limbs and head are more masculine than feminine due to the early masculinization of my skeleton. However, as I have finally fully 'matured,' my body has begun to 'break down,' and the somewhat masculine aspects of my physiology are too. My arms have been the first to feel this, as they are in a state of constant pain due to their 'excessive' use/abuse in the physically demanding environment I work in. I don't know how much longer my body will be able to take this punishment.

Had I been 'left alone,' I would have developed as a female with a small penis. There may have been a small amount of male development, but it cannot be known as to how much. It is my suspicion that I would have first gone through female puberty at about 9 or so, and later, at about 14 or so, I might have gone through male puberty. However, the advent of female puberty may have delayed or eliminated the maturation of the testicle as estriadol production would have caused the cessation of spermatogenesis. This is the primary reason that 'true' hermaphrodites rarely sire children. Only three have been documented - this fact is borne out through documentation on "ovo-testis" that I provided a number of months back.

I guess that makes me number four. But, that is due solely to the surgical intervention of the butchers who did this to me. What is most horrific about my experience, more than anything else, is callous, brutal, arrogance of the so-called 'doctors' who used deceitful and immoral, if not illegal, means to ensure that they got to play doctor, and perform their experimental procedures on a "bastard hermaphrodite child being put up for adoption." They told my adoptive parents that I was a "Perfect Specimen." What they meant was that I was the "Perfect human guinea pig" – the Perfect Lab Rat.

I'm now going through counseling, but it remains to be seen what, if any, help it is. The psychiatrist is intelligent and thoughtful but, just the same, is something of the 'old school' line of thought. It seems to me that he's just following the established protocols – i.e.; keep me in the 'male' camp, but letting me 'experiment' with my 'female' side – just don't take it too far!

What else is there to say? The emotional scarring from the brainwashing and lies required to enforce my male assignment will never truly heal. I still have all the female 'programming and wiring,' along with a small portion of my 'plumbing' that I was born with, but my body has been horribly mutilated and mutated into something it was never meant to be. I don't hold out a whole lot of hope for the future. I will never be able to feel 'whole' – too much was taken away when I was an infant, and even more was destroyed during my upbringing.

Anyway, I just had to let you all know what I was dealing with, and thought it would be good for our community to know one more 'story' of the how the medical community treats 'us.'

Forewarned is forearmed.

All my best to all of you.

Sharon's comments are illustrative of the other side of the story. Because she can witness online to others like her, and because still others in the general public can find and read that confession, Sharon's words not only are cathartic for her, but they are also counter-discursive. The "medical community" is not for her an authority she goes to for help, but is rather an adversary that, she writes, "treats 'us,'" as illustrated through her experience, using an " 'old school' line of thought."

On November 9, 2004, Beatrice responds, supportive and understanding of Sharon's experience:

> Hi…wow your story is interesting , and your surgerys [*sic*] and scar placement ,are so different [*sic*] than mine , i was 13 when my surgerys [*sic*] were done , and the delayed puberty is interesting , hitt [*sic*] me all at once , if you want to talk , im [*sic*] here.

## Intersex Geeks

Intersex virtual neighborhoods demonstrate that there are clearly those living outside the boundaries of perceived normality. Where little research exists within medical literature to document the impact of past treatment practices (Reiner, "Mixed Method" 51), the Internet is available to provide countless documentation, revealed through narratives told from the intersex person's perspective. "It's no surprise," Koyoma states, "that [the] intersex movement uses [the] Internet as a tool to reach its constituents and to educate the public" (personal communication, March 25, 2005). Many of the leading figures in the intersex movement, including Koyoma and ISNA founder Cheryl Chase, are self-proclaimed geeks. Chase is, Dreger writes, an "expert computer programmer and software developer" ("Question"). In fact, she ran a software company in Japan before founding ISNA, and now, while she is administering day-to-day operations at ISNA, she also is working on a content management system specifically designed for nonprofit organizations. Koyoma, who once interned at ISNA and helped to develop one of the versions of its Web site before developing the Intersex Initiative site, began programming when she was in elementary school, and later in life wrote computer software for pay ("Question").

The intersex movement is also one of the first to start after the Internet Revolution, so leaders of the movement already adept at using technology quickly jumped at the opportunity to exploit it to enhance the effectiveness of their efforts. As a result, many intersex Web sites, such as those of ISNA, Intersex Initiative, and also BLO, display a level of technical sophistication that both gives them an authoritative, professional look and also enables them to exploit available technologies to their fullest in an effort to facilitate communications among site visitors or push information to the media and public.

Preves asserts that the Internet "already has afforded the development of a geographically diverse advocacy movement" (*Intersex* 95). The Internet, more than any other technology available, provides to intersex persons an opportunity to perform an identity otherwise denied to them. Even if it is a virtual place, and even if it has particular problems, the Internet is still a safe world of empowerment, one that offers anonymity and, at the same time, offers the chance to commune with others. Online, Sherry Turkle writes, "We are encouraged to think of ourselves as fluid, emergent, decentralized, multiplicitous, flexible, and ever in process" (263–264).

Technology is becoming cheaper and so access to computers and the Internet is becoming possible for greater numbers of previously technologically isolated persons. Web hosting services, for example, which provide intersex Web sites with server space and, in some cases, administration help in maintaining an Internet presence, cost only dollars each month. Some ISPs (Internet Service Providers) even offer free Web site hosting, so an intersex virtual community can literally be built and kept active at little or no cost. This is especially true if volunteers contribute time to keeping up the Web site. So intersex activists who might otherwise never meet or share information can virtually, and cheaply, engage in community-building and counter-discourse. As Sandor Vegh points out, "[O]nly the Internet allows an activist to distribute a message to thousands of people all over the world at once and to publish information that is accessible from anywhere anytime with virtually no cost" (74).

Once the domain of only the affluent or educated, the Internet now, because of its increasing affordability, is much more available and consequently is used more often by a wide range of people. Even owning a personal computer is no longer a requirement for going online, as a variety of other electronic devices, including a cell phone, allow for Internet connectivity that is not even tied to a particular spot. One can, because of advances in wireless technology, cell phones, and other portable electronic devices, access the Internet from anywhere at any time. And even when the lowered costs of computers and other technologies are not realistic for some, public libraries and other places offer free computer and Internet access.

## Final Thoughts

The Internet represents for intersex activists a legitimate opportunity for challenging, Howard Rheingold writes, "the existing political hierarchy's monopoly" (14) on power. Those intersex persons who might not have the financial means or the privileged status to communicate their ideas and positions on the subject of intersexuality can do so because of the Internet. In addition, those intersex persons whose opinions have been ignored or not privileged enough to be considered legitimate can be influential because of the Internet. As Driver writes, "[F]or many years there was a misperception that only a few intersex people were upset over the treatment they received." The Intersex virtual community "illustrates otherwise." Virtual intersex neighborhoods, in fact, show that intersex persons are more common than what we have been told, they have stories to tell that we have not been told, and they have alternative ideas and suggestions for treatment, or for what it means to be a man or a woman, that challenge directly the truths we have been told. When, before, few means have existed outside the classroom, traditional media, or the medical clinic to learn of such things, the Internet now empowers their telling, and in that telling the noise of cultural reproduction can be heard, loudly churning out communities, information, and resistance.

# References


American Academy of Pediatrics. "Evaluation of the Newborn with Developmental Anomalies of the External Genitalia." *Pediatrics* 106.1 (2000): 138–142. Print.

*American Association for Klinefelter Syndrome Information and Support.* n.d. Web. 10 Aug. 2004.

*Androgen Insensitivity Syndrome Support Group.* n.d. Aissg.org. Web. 10 Aug. 2004.

"Apologetic." *Intersex Support Group International: ISGI.* n.d. Xyxo.org. Web. 30 June 2003.

Appadurai, Arjun. *Modernity at Large: Cultural Dimensions of Globalization.* Minneapolis: University of Minnesota Press, 1996.

Berland, G. K., et al. "Health Information on the Internet: Accessibility, Quality, and Readability in English and Spanish." *Journal of the American Medical Association* 285.20 (2001): 2612–2621. Print.

Biesecker, Barbara. "Michel Foucault and the Question of Rhetoric." *Philosophy and Rhetoric* 25 (1992): 351–364. Print.

*Bodies Like Ours.* n.d. Bodieslikeours.org. Web. 10 Aug. 2004.

Boyd, Danah M., and Nicole B. Ellison. "Social Network Sites: Definition, History, and Scholarship." *Journal of Computer-Mediated Communication* 13.1 (2007). Jcmc.indiana.edu. Web. 8 May 2008.

Butler, Judith. *Bodies That Matter: On the Discursive Limits of Sex*. New York: Routledge, 1993. Print.

———. *Gender Trouble: Feminism and the Subversion of Identity.* New York: Routledge, 1990. Print.

———. *Undoing Gender*. New York: Routledge, 2004. Print.

Carter, Denise. "Living in Virtual Communities: An Ethnography of Human Relationships in Cyberspace." *Information, Communication & Society* 8.2 (June 2005): 148–167. Print.

Chase, Cheryl. "Research Request Regarding the ISNA Web Site." Message to the author. 4 August 2004. E-mail.

———. "Surgical Progress Is Not the Answer to Intersexuality." *Intersex and the Age of Ethics.* Ed. Alice Domurat Dreger. Hagerstown, MD: University Publishing Group, 1999. 147–159. Print.

———. "What is the Agenda of the Intersex Patient Advocacy Movement?" *Endocrinologist* 13.3 (2003): 240–242. Print.

Code, Lorraine. *Rhetorical Spaces: Essays on Gendered Locations.* New York: Routledge, 1995. Print.

Colapinto, John. *As Nature Made Him: The Boy Who Was Raised as a Girl.* New York: Harper Collins, 2001. Print.

Considine, Bob. "Woman Spots Baby's Eye Cancer Online." *MSNBC Interactive*. 28 Aug. 2008. Msnbc.msn.com. Web. 29 Aug. 2008.

Corpron, Cynthia, and Joseph L. Lelli, Jr. "Evaluation of Pediatric Surgery Information on the Internet." *Journal of Pediatric Surgery* 36.8 (2001): 1187–1189. Print.

Crouch, Robert A. "Betwixt and Between: The Past and Future of Intersexuality." *Intersex and the Age of Ethics.* Ed. Alice

Domurat Dreger. Hagerstown, MD: University Publishing Group, 1999. 29–49. Print.

Daaboul, Jorge, and Joel Frader. "Ethics and the Management of the Patient with Intersex: A Middle Way." *Journal of Pediatric Endocrinology & Metabolism* 14 (2001): 1575–1583. Print.

Damiani, Durval. "Ambiguous Terms Still Exist in the Consensus." *Electronic Letters to Hughes et al's "Consensus Statement on Management of Intersex Disorders."* 19 July 2006. Web. 4 April 2007. http://adc.bmj.com/cgi/eletters/91/7/554.

Dartnell, Michael V. *Insurgency Online: Web Activism and Global Conflict.* Toronto: University of Toronto Press, 2003. Print.

"Dave and Christine's Story." *Androgen Insensitivity Syndrome Support Group.* Aug. 2003. Aissg.org. Web. 6 Nov. 2004.

Dean, Carolyn J. "The Productive Hypothesis: Foucault, Gender, and the History of Sexuality." *History and Theory* 33.3 (1994): 271–296. Print.

Deleuze, Gilles, and Felix Guattari. *A Thousand Plateaus: Capitalism and Schizophrenia.* Minnesota: University of Minnesota Press, 1987. Print.

Diamond, Milton, and Hazel G. Beh. "Variations of Sex Development Instead of Disorders of Sex Development." *Electronic Letters to Hughes et al's "Consensus Statement on Management of Intersex Disorders."* 2 Aug. 2006. Web. 4 April 2007. http://adc.bmj.com/cgi/eletters/91/7/554

Diamond, Milton, and Keith Sigmundson. "Management of Intersexuality: Guidelines for Dealing with Persons of Ambiguous Genitalia." *Archives of Pediatrics & Adolescent Medicine* 151.10 (1997): 1046–1050. Print.

Dreger, Alice Domurat. "Ambiguous Sex—or Ambivalent Medicine? Ethical Issues in the Treatment of Intersexuality." *The Hastings Center Report* 28.3 (1998): 24–35. Print.

———. *Hermaphrodites and the Medical Invention of Sex*. Cambridge, MA: Harvard University Press, 1998. Print.

———, ed. *Intersex and the Age of Ethics*. Hagerstown, MD: University Publishing Group, 1999. Print.

———. "Question Regarding the Importance of the Internet for Intersex Activism." Message to the author. 23 March 2005. E-mail.

Driver, Betsy. "Research Request Regarding the Bodies Like Ours Web Site." Message to the author. 2 Aug. 2004. E-mail.

Du Gay, Paul, et al. *Doing Cultural Studies: The Story of the Sony Walkman*. London: Sage, 1997. Print.

Ensler, Eve. *The Vagina Monologues*. New York: Villard, 1998. Print.

Eugenides, Jeffrey. *Middlesex*. New York: Farrar, Straus, and Giroux, 2002. Print.

Fausto-Sterling, Anne. *Sexing the Body: Gender Politics and the Construction of Sexuality*. New York: Basic Books, 2000. Print.

Foucault, Michel. *The Archaeology of Knowledge and the Discourse on Language*. Trans. A. M. Sheridan Smith. New York: Pantheon Books, 1972. Print.

———. *The Birth of the Clinic: An Archaeology of Medical Perception*. Trans. A.M. Sheridan Smith. New York: Vintage Books, 1994. Print.

———. *Discipline and Punish: The Birth of the Prison*. Trans. Alan Sheridan. New York: Vintage Books, 1995. Print.

———. *Herculine Barbin: Being the Recently Discovered Memoirs of a Nineteenth Century French Hermaphrodite.* Trans. Richard McDougall. New York: Pantheon Books, 1980. Print.

———. *The History of Sexuality: An Introduction, Volume 1.* Trans. Robert Hurley. New York: Vintage Books, 1990. Print.

———. *The Order of Things: An Archaeology of the Human Sciences*. New York: Pantheon, 1970. Print.

———. *Power/Knowledge: Selected Interviews and other Writings, 1972–1977.* Ed. C. Gordon, et al. Brighton, CT: Harvester, 1980. Print.

———. "What is Critique?" *The Political.* Ed. David Ingram. Malden, MA: Blackwell, 2002: 191–211. Print.

Friedman, Batya, Peter H. Kahn, Jr., and Daniel C. Howe. "Trust Online." *Communications of the ACM* 43.12 (December, 2000): 34–40. Print.

Gayle. "Gayle's Story." *Androgen Insensitivity Syndrome Support Group (AISSG)*. Oct. 2002. Aissg.org. Web. 9 May 2007.

Glogoff, Stuart. "Virtual Connections: Community Bonding on the Net." *First Monday* 6.3 (2001). Web. 6 February 2007.

Groveman, Sheri A. "The Hanukkah Bush: Ethical Implications in the Clinical Management of Intersex." *Intersex and the Age of Ethics.* Ed. Alice Domurat Dreger. Hagerstown, MD: University Publishing Group, 1999: 23–28. Print.

Hall, Stuart. "Introduction." *Representation: Cultural Representations and Signifying Practices.* Ed. Stuart Hall. London: Sage, 1997: 1–14. Print.

———. "Introduction: Who Needs 'Identity'?" Questions of Cultural Identity. Ed. Stuart Hall and Paul du Gay. London: Sage, 1997: 1–17. Print.

———. "The Local and the Global: Globalization and Ethnicity." *Culture, Globalization and the World-System.* Ed. Anthony King. Minneapolis: University of Minnesota Press, 1997: 19–40. Print.

Hartman, Debbie. *Personal Web Site.* n.d. Geocities.com. Web. 30 Mar. 2005. http://www.geocities.com/elee2_2000/indexdebandkellspage.html

Hawkins, Anne Hunsaker. *Reconstructing Illness: Studies in Pathography.* West Lafayette, Indiana: Purdue University Press, 1999. Print.

Hendricks, Melissa. "Into the Hands of Babes." *Johns Hopkins University Magazine.* Sep. 2000. Jhu.edu. Web. 23 April 2004.

Herdt, Gilbert, ed. *Third Sex, Third Gender: Beyond Sexual Dimorphism in Culture and History.* New York: Zone Books, 1994. Print.

———. "Introduction: Third Sexes and Third Genders." *Third Sex, Third Gender: Beyond Sexual Dimorphism in Culture and History.* Ed. Gilbert Herdt. New York: Zone Books, 1994. 21–81. Print.

———. "Mistaken Sex: Culture, Biology and the Third Sex in New Guinea. *Third Sex, Third Gender: Beyond Sexual Dimorphism in Culture and History.* Ed. Gilbert Herdt. New York: Zone Books, 1994. 419–445. Print.

*Hermaphrodite Education and Listening Post (HELP).* n.d. Jax-inter.net. Web. 27 July 2004.

Hinkle, Curtis. "Deconstructing the Feminist Essence Narrative: Ongoing Investigation by OII." *Organisation Internationale des Intersexes.* 20 April 2008. Intersexualite.org. Web. 17 July 2008.

Holmes, Morgan. "Rethinking the Meaning and Management of Intersexuality." *Sexualities* 5.2 (2002): 159–180. Print.

Howard, Philip N., and Steve Jones, eds. *Society Online: The Internet in Context.* Thousand Oaks, CA: Sage Publishing, 2004. Print.

Howe, Edmund G. "Advances in Treating (or Not Treating) Intersexed Persons: Understanding Resistance to Change. *Ethics and Intersex*. Ed. Sharon E. Sytsma. Netherlands: Springer Dordrecht, 2006: 115–137. Print.

Hughes, I. A., et al. "Consensus Statement on Management of Intersex Disorders." *Arch Dis Child* 91 (2006): 554–562. Print.

*Intersex Collective.* n.d. Intersexcollective.org. Web. 11 Aug. 2008.

"Internet Growth Statistics." *Internet World Stats: Usage and Population Statistics.* 18 Sep. 2008. Internetworldstats.com. Web. 18 Sep. 2008.

*Intersex Initiative*. n.d. Intersexinitiative.org. Web. 10 Aug. 2004.

*Intersex Society of North America.* n.d. Isna.org. Web. 26 June 2003.

———. "Guiding Tips for Parents." *Intersex Society of North America (ISNA).* n.d. Isna.org. Web. 2 April 2005.

———. "ISNA Outreach for 2004." 2004. Isna.org. Web. 26 March 2005.

———. "PBS Airs Documentary About Intersex Without Any Intersex People in It." 2001. Isna.org. Web. 19 Aug. 2008.

*Intersex Support Group International.* n.d. Xyxo.org. Web.18 June 2003.

Kessler, Suzanne J. *Lessons from the Intersexed.* New Brunswick, NJ: Rutgers University Press, 2002. Print.

Kilty, Sharon. "Telling the Illness Story." *The Patients' Network* 5.3 (Winter 2000). Patient-network.com. Web. 14 May 2007.

Kipnis, Kenneth, and Milton Diamond. "Pediatric Ethics and the Surgical Assignment of Sex." *Intersex and the Age of Ethics*. Ed. Alice Domurat Dreger. Hagerstown, MD: University Publishing Group, 1999. 173–193. Print.

*Klinefelter Syndrome Support Group.* n.d. Klinefeltersyndrome.org. Web. 10 Aug. 2004.

Koerber, Amy. "Postmodernism, Resistance, and Cyberspace: Making Rhetorical Spaces for Feminist Mothers on the Web." *Women's Studies in Communication* 24.2 (2001): 218–240. Print.

Koyoma, Emi. "From 'Intersex' to 'DSD:' Toward a Queer Disability Politics of Gender." *Translating Identity Conference at University of Vermont* Feb. 2006. Uvm.edu. Web. 16 April 2007.

———. "Question Regarding the Importance of Internet for Intersex Activism." Message to the author. 25 March 2005. E-mail.

Kuhn, Thomas S. *The Structure of Scientific Revolutions.* 3rd ed. Chicago: University of Chicago Press, 1996. Print.

Laquer, Thomas. *Making Sex: Body and Gender from the Greeks to Freud.* Cambridge, MA: Harvard University Press, 1990. Print.

Lee, Peter A., et al. "Consensus Statement on Management of Intersex Disorders." *Pediatrics* 118.2 (2006): 488–500. Print.

Long, Lynnell Stephani. "DSD vs Intersex." *Electronic Letters to Hughes et al's "Consensus Statement on Management of Intersex Disorders."* 23 Aug. 2006. Web. 4 April 2007. http://adc.bmj.com/cgi/eletters/91/7/554

———. "Question Regarding the Importance of Internet for Intersex Activism." Message to the author. 28 March 2005. E-mail.

Matei, Sorin Adam. "A Sounding Board for the Self: Virtual Community as Ideology." *Journal of Technical Writing and Communication* 35.4 (2005): 345–365. Print.

———. "From Counterculture to Cyberculture: Virtual Community Discourse and the Dilemma of Modernity." *Journal of Computer-Mediated Communication* 10.3 (2005). Jcmc.indiana.edu. Web. 16 May 2006.

Matta, Christina. "Ambiguous Bodies and Deviant Sexualities: Hermaphrodites, Homosexuality, and Surgery in the United States, 1850–1904." *Perspectives in Biology and Medicine* 48.1 (Winter, 2005): 74–83. Print.

Money, John. *Sex Errors of the Body: Dilemmas, Education, Counseling*. Baltimore: Johns Hopkins University Press, 1968. Print.

Money, John, and Patricia Tucker. *Sexual Signatures.* Boston: Little, Brown and Company, 1975. Print.

Morland, Iain. "Postmodern Intersex." *Ethics and Intersex.* Ed. Sharon E. Sytsma. Netherlands: Springer Dordrecht, 2006: 319–332. Print.

Morris, Esther. "The Missing Vagina Dialogue." 2004. Web. 3 Jan. 2004. http://mrkh.org/files/The_Missing_Vagina_Monologue_and_Beyond.pdf

*MRKH.org: A Resource for Women with Mayer Rokitansky Kuster Hauser Syndrome*. n.d. Mrkh.org. Web. 22 Oct. 2004.

"MRKHDocs: Recommended Doctors and Therapists." *MRKH.org*. n.d. Mrkh.org. Web. 21 February 2005.

Orgad, Shani. "The Cultural Dimensions of Online Communication: A Study of Breast Cancer Patients' Internet Spaces." *New Media & Society* 8.6 (2006): 877–899. Print.

*OII—USA (Organisation Intersex International)*. n.d. Intersexualite.org. Web. 23 April 2007.

*Organisation Internationale des Intersexués*. n.d. Intersexualite.org. Web. 23 April 2007

———. "No Normalisation Zone." n.d. Intersexualite.org. Web. 14 May 2008.

Porter, Constance Elise. "A Typology of Virtual Communities: A Multi-Disciplinary Foundation for Future Research." *Journal of Computer-Mediated Communication* 10.1 (2004). Jcmc.indiana.edu. Web. 22 Dec. 2004.

Preves, Sharon E. *Intersex and Identity: The Contested Self*. New Brunswick, NJ: Rutgers University Press, 2003. Print.

———. "Out of the O.R. and into the Streets: Exploring the Impact of Intersex Media Activism." *Research in Political Sociology* 13 (2004): 179–223. Print.

"Priestly Sins." (1992, October 7). *Oprah: The Oprah Winfrey Show*. Transcript. Livingston, NJ: Burrelle's Information Services.

Radin, Patricia. "'To Me, It's My Life': Medical Communication, Trust, and Activism in Cyberspace." *Social Science & Medicine* 62 (2006): 591–601. Print.

Reiner, William G. "Gender Identity and Sex Assignment: A Reappraisal for the 21st Century." *Pediatric Gender Assignment: A Critical Reappraisal.* Ed. Stephen Zderic, et al. New York: Kluwer Academic/Plenum Publishers, 2002: 175–189. Print.

———. "Mixed Method Research for Child Outcomes in Intersex Conditions." *BJU International* 93 (2004): 51–53. Print.

Reis, Elizabeth. "Impossible Hermaphrodites: Intersex in America, 1620–1960." *The Journal of American History* 92.2 (September 2005). Journalofamericanhistory.org. Web. June 25, 2007.

Reizenberger, Azila Talit. "The Creation of Adam as Hermaphrodite—and Its Implications for Feminist Theology." *Judaism* (Fall 2003). Print.

Rheingold, Howard. *The Virtual Community: Homesteading on the Electronic Frontier*. Reading, MA: Addison-Wesley Publishing Co., 1993. Print.

Ridings, Catherine M., and David Gefen. "Virtual Community Attraction: Why People Hang Out Online." *Journal of Computer-Mediated Communication* 10.1 (Nov. 2004). Jcmc.indiana.edu. Web. 23 Jan. 2006.

Ridings, Catherine M., David Gefen, and Bay Arinze. "Some Antecedents and Effects of Trust in Virtual Communities." *Journal of Strategic Information Systems* 11 (2002): 271–295. Print.

Segal, Judy Z. *Health and the Rhetoric of Medicine.* Carbondale and Edwardsville: Southern Illinois University Press, 2005. Print.

Shakespeare, William. *Henry IV. The Riverside Shakespeare.* Ed. G. Blakemore Evans and J. J. M. Tobin. Boston: Houghton Mifflin Co., 1974. 842–929. Print.

Sherman, Lori D. "Diagnosis and Management of Ambiguous Genitalia." *Indian Journal of Pediatrics* 64 (1997): 195–203. Print.

Stark, Herman E. "Authenticity and Intersexuality." *Ethics and Intersex.* Ed. Sharon E. Sytsma. Netherlands: Springer Dordrecht, 2006: 271–291. Print.

Sytsma, Sharon E., ed. *Ethics and Intersex.* Netherlands: Springer Dordrecht, 2006. Print.

Turkle, Sherry. *Life on the Screen: Identity in the Age of the Internet.* New York: Simon and Schuster, 1995. Print.

*Turner Syndrome Society of the United States.* n.d. Turnersyndrome.org. Web. 10 Aug. 2004.

*UK Intersex Association.* n.d. Ukia.co.uk. Web. 10 Aug. 2004.

Vegh, Sandor. "Classifying Forms of Online Activism: The Case of Cyberprotests Against the World Bank." *Cyberactivism: Online Activism in Theory and Practice.* Ed. Martha McCaughey and Michael D. Ayers. New York: Routledge, 2003. 71–95. Print.

"Web Sites: Number of Pages." *Digital Design Works.* n.d. Web. 3 July 2008. http://homepage.mac.com/colleenmiller/ddw/docs/pages.html.

Weiler, R. M., and Pealer, L. N. "The Sitelegend: Twelve Components of a New Strategy for Providing Website Documentation." *Journal of School Health* 70.4 (2000): 148–152. Print.

Wellman, Barry, and Milena Gulia. "Net Surfers Don't Ride Alone: Virtual Communities as Communities." Department of Sociology and Centre for Urban and Community Studies. University of Toronto. Aug. 1997: 1–26. Utoronto.ca. Web.14 Aug. 2005.

"WWW Size." 14 June 2008. Worldwidewebsize.com. Web. 14 June 2008. <http://www.worldwidewebsize.com/>.

Zakon, Robert. *Hobbes Internet Timeline* vol. 8.2. 1 Nov. 2006. Zakon.org. Web. 12 June 2008.

Zucker, Kenneth J. "Gender Identity and Intersexuality." *Ethics and Intersex*. Ed. Sharon E. Sytsma. Netherlands: Springer Dordrecht, 2006: 165–181. Print.

# Index

Made in the USA
Monee, IL
12 November 2026

42a60044-4ec3-45b1-9af9-c2edf70f1b52R01